TECHNICAL ISSUES RELATED TO THE COMPREHENSIVE NUCLEAR TEST BAN TREATY

Committee on Technical Issues Related to Ratification of the
Comprehensive Nuclear Test Ban Treaty

NATIONAL ACADEMY OF SCIENCES

NATIONAL ACADEMY PRESS
Washington, D.C.

NATIONAL ACADEMY PRESS   2101 Constitution Avenue, N.W.   Washington, D.C. 20418

NOTICE: The project that is the subject of this report was approved by the Council of the National Academy of Sciences. The members of the panel responsible for the report were chosen for their special competences and with regard for appropriate balance.

This study was supported by Contract No. DE-AM01-99PO90016, Task Order No. DE-AT01-01NN40245 between the National Academy of Sciences and the Department of Energy, Contract No. S-LMAQM-00-C-0125 between the National Academy of Sciences and the Department of State, the William and Flora Hewlett Foundation, the Carnegie Corporation of New York, the John D. and Catherine T. MacArthur Foundation, and National Research Council Funds. Any opinions, findings, conclusions, or recommendations expressed in this publication are those of the author(s) and do not necessarily reflect the views of the organizations or agencies that provided support for the project.

International Standard Book Number 0-309-08506-3

Additional copies of this report are available from the Committee on International Security and Arms Control, 2101 Constitution Avenue, N.W., Washington, D.C. 20418, (202) 334-2811, cisac@nas.edu; the report is also available online at http://www.nap.edu.

Printed in the United States of America
Copyright 2002 by the National Academy of Sciences. All rights reserved.

THE NATIONAL ACADEMIES

National Academy of Sciences
National Academy of Engineering
Institute of Medicine
National Research Council

The **National Academy of Sciences** is a private, nonprofit, self-perpetuating society of distinguished scholars engaged in scientific and engineering research, dedicated to the furtherance of science and technology and to their use for the general welfare. Upon the authority of the charter granted to it by the Congress in 1863, the Academy has a mandate that requires it to advise the federal government on scientific and technical matters. Dr. Bruce M. Alberts is president of the National Academy of Sciences.

The **National Academy of Engineering** was established in 1964, under the charter of the National Academy of Sciences, as a parallel organization of outstanding engineers. It is autonomous in its administration and in the selection of its members, sharing with the National Academy of Sciences the responsibility for advising the federal government. The National Academy of Engineering also sponsors engineering programs aimed at meeting national needs, encourages education and research, and recognizes the superior achievements of engineers. Dr. Wm. A. Wulf is president of the National Academy of Engineering.

The **Institute of Medicine** was established in 1970 by the National Academy of Sciences to secure the services of eminent members of appropriate professions in the examination of policy matters pertaining to the health of the public. The Institute acts under the responsibility given to the National Academy of Sciences by its congressional charter to be an adviser to the federal government and, upon its own initiative, to identify issues of medical care, research, and education. Dr. Harvey V. Fineberg is president of the Institute of Medicine.

The **National Research Council** was organized by the National Academy of Sciences in 1916 to associate the broad community of science and technology with the Academy's purposes of furthering knowledge and advising the federal government. Functioning in accordance with general policies determined by the Academy, the Council has become the principal operating agency of both the National Academy of Sciences and the National Academy of Engineering in providing services to the government, the public, and the scientific and engineering communities. The Council is administered jointly by both Academies and the Institute of Medicine. Dr. Bruce M. Alberts and Dr. Wm. A. Wulf are chairman and vice chairman, respectively, of the National Research Council.

COMMITTEE ON TECHNICAL ISSUES RELATED TO RATIFICATION OF THE COMPREHENSIVE TEST BAN TREATY

JOHN P. HOLDREN (*Chair*), Director, Program in Science, Technology & Public Policy, John F. Kennedy School of Government, Harvard University, Cambridge, Massachusetts, and Chair, Committee on International Security and Arms Control, National Academy of Sciences

HAROLD AGNEW, President (retired), General Atomics; Director (retired), Los Alamos National Laboratory, Los Alamos, New Mexico

RICHARD L. GARWIN, Senior Fellow for Science and Technology, Council on Foreign Relations, New York, New York; Emeritus Fellow, Thomas J. Watson Research Center, IBM Corporation

RAYMOND JEANLOZ, Professor, Department of Earth and Planetary Science, and Professor, Department of Astronomy, University of California at Berkeley, Berkeley, California; Member, National Security Panel, University of California President's Council

SPURGEON M. KEENY, JR., Senior Fellow, National Academy of Sciences; President (retired), Arms Control Association, Washington D.C.; and Deputy Director (retired) U.S. Arms Control and Disarmament Agency

CHARLES LARSON, Admiral (USN, Ret.); former Commander in Chief of the Unified Pacific Command; Superintendent, U.S. Naval Academy, Annapolis, Maryland

ALBERT NARATH, Director, (retired), Sandia National Laboratories, Albuquerque, New Mexico

WOLFGANG K.H. PANOFSKY, Professor and Director Emeritus, Stanford Linear Accelerator Center, Stanford University, Stanford, California

PAUL G. RICHARDS, Mellon Professor of Natural Science, Lamont-Doherty Earth Observatory, Columbia University, Palisades, New York

SEYMOUR SACK, Laboratory Associate, Lawrence Livermore National Laboratory, Livermore, California

ALVIN W. TRIVELPIECE, President (retired), Lockheed Martin Energy Research Corporation; Director (retired), Oak Ridge National Laboratory, Oak Ridge, Tennessee

Study Staff
JO L. HUSBANDS, Study Director
DAVID HAFEMEISTER, Staff Officer
CHRISTOPHER ELDRIDGE, Staff Officer
LA'FAYE LEWIS-OLIVER, Financial Associate
AMY GIAMIS, Program Assistant

COMMITTEE ON INTERNATIONAL SECURITY AND ARMS CONTROL

JOHN P. HOLDREN (*Chair*), Director, Program in Science, Technology & Public Policy, John F. Kennedy School of Government, Harvard University, Cambridge, Massachusetts, and Chair, Committee on International Security and Arms Control, National Academy of Sciences

CATHERINE KELLEHER (*Vice-Chair*), Visiting Professor, Naval War College, Newport, Rhode Island

JOHN D. STEINBRUNER *(Vice-Chair)*, Professor and Director, Center for International and Security Studies at Maryland, School of Public Affairs, University of Maryland, College Park, Maryland

WILLIAM F. BURNS, Major General (USA, Ret.), Carlisle, Pennsylvania

GEORGE LEE BUTLER, President, Second Chance Foundation, Omaha, Nebraska*

STEPHEN COHEN, Senior Fellow, Foreign Policy Studies Program, The Brookings Institution, Washington, D.C.

SUSAN EISENHOWER, The Eisenhower Institute, Washington D.C.

STEVE FETTER, School of Public Affairs, University of Maryland, College Park, Maryland

ALEXANDER H. FLAX, President Emeritus, Institute for Defense Analyses, and Senior Fellow, National Academy of Engineering, Washington D.C.

RICHARD L. GARWIN, Senior Fellow for Science and Technology, Council on Foreign Relations, New York, New York; Emeritus Fellow, Thomas J. Watson Research Center, IBM Corporation

SPURGEON M. KEENY, JR., Senior Fellow, National Academy of Sciences; President (retired), Arms Control Association, Washington D.C.; and Deputy Director (retired) U.S. Arms Control and Disarmament Agency

CHARLES LARSON, Admiral (USN, Ret.) U.S. Naval Academy, Annapolis, Maryland**

JOSHUA LEDERBERG, University Professor, The Rockefeller University, New York, New York

MATTHEW MESELSON, Thomas Dudley Cabot Professor of the Natural Sciences, Department of Molecular and Cellular Biology, Harvard University, Cambridge, Massachusetts

ALBERT NARATH, Director, (retired), Sandia National Laboratories, Albuquerque, New Mexico

WOLFGANG K.H. PANOFSKY, Professor and Director Emeritus, Stanford Linear Accelerator Center, Stanford University, Stanford, California

C. KUMAR N. PATEL, Professor, Department of Physics and Astronomy, University of California, Los Angeles

JONATHAN D. POLLACK, Professor of Asian and Pacific Studies and Director, Strategic Research, Naval War College, Newport, Rhode Island

F. SHERWOOD ROWLAND, ex officio, Foreign Secretary, National Academy of Sciences, Washington D.C.

Committee Staff
JO L. HUSBANDS, Director
DAVID HAFEMEISTER, Staff Officer
PATRICIA STEIN WRIGHTSON, Staff Officer
EILEEN CHOFFNES, Staff Officer
CHRISTOPHER ELDRIDGE, Staff Officer
LA'FAYE LEWIS-OLIVER, Financial Associate
AMY GIAMIS, Program Assistant

* Until March 3, 2001.

** Until April 5, 2002.

PREFACE

Since the beginning of the nuclear era, the international community has debated proposals to reduce the risks posed by the existence of nuclear weapons and their proliferation. The environmental hazards of nuclear test explosions in the atmosphere, added to the dangers inherent in the nuclear arms competition, led to early initiatives designed to limit nuclear testing. In 1958 and 1959 groups of Soviet and American scientists met to discuss the technical issues raised by a potential ban on all nuclear tests. As a leading seismologist, Frank Press, my predecessor as President of the National Academy of Sciences (NAS), was heavily engaged in the discussions within the U.S. scientific community over the desirability and technical feasibility of a comprehensive test ban, and he participated in the international dialogue. These issues have continued to engage U.S. scientists ever since.

The study that follows here resulted from a request to the NAS in April 2000 from General John Shalikashvili, (U.S. Army, ret.), then the Special Advisor to the President and the Secretary of State for the Comprehensive Test Ban Treaty (CTBT). General Shalikashvili had been asked, after the U.S. Senate voted against providing its advice and consent to the ratification of the CTBT, to examine the major technical and political concerns that had led to the Senate's rejection of the treaty and to explore a possible basis for its reconsideration. To support his efforts, General Shalikashvili commissioned several studies, including this one from the Academy, to address the major technical issues that had arisen during the Senate debate.

The Academy study was not asked to provide an overall "net assessment" of whether the CTBT is in the national security interest of the United States, and it did not do so. Its mandate was confined, rather, to a specified set of important technical questions that, along with political questions and other technical ones we were not asked to address, are relevant to this larger issue.

The formal U.S. government sponsor of the NAS study was the Department of State, with funding provided by the Department of Energy. Additional support for the study was provided by the John D. and Catherine T. MacArthur Foundation, the Carnegie Corporation of New York, the William and Flora Hewlett Foundation, and internal funds of The National Academies.

To organize the study, the NAS turned to its standing Committee on International Security and Arms Control (CISAC), which was created in 1980 to bring the scientific and technical resources of the NAS to bear on critical security issues. CISAC conducts policy studies and carries out a program of private, off-the-record dialogues with counterpart groups in Russia, China, and India. CISAC worked with the NAS leadership and government sponsors to define the scope of the study. The committee that I appointed to carry out the study—the Committee on Technical Issues Related to Ratification of the Comprehensive Nuclear Test Ban Treaty (the CTBT Committee) contains a number of CISAC members, including CISAC chair John P. Holdren (Teresa and John Heinz Professor of Environmental Policy, John F. Kennedy School of Government, Harvard University) and CISAC chair emeritus Wolfgang K.H. Panofsky (Professor and Director Emeritus, Stanford Linear Accelerator Center), but has operated independently from CISAC. The report was written by the CTBT Committee and reviewed through the usual Academy process (see the Acknowledgments); those members of CISAC not on the CTBT Committee did not review the report and are not responsible for its contents.

The CTBT committee began its work on July 1, 2000 and undertook a demanding schedule of briefings and meetings in order to be able to provide a meaningful progress report to General Shalikashivili before completion of his report to the President and Secretary of State in January 2001.[1] The committee had extensive access to classified reports and material and held all but its first meeting at the classified level. After the committee signed off on its draft report in December 2000, the report then entered what became an extended process of multi-agency classification review, Academy peer review, and classification re-review after modifications made in response to the Academy review. This long process, which was finally completed in June 2002, was necessary to meet the combined requirements of classification rules and Academy rigor. Although it delayed the final report beyond what anyone had expected, in both the committee's judgment and mine the findings remain both current and highly relevant to the current policy context.

Several members of The National Academies staff contributed significantly to the preparation and production of the report. CISAC's staff director, Jo Husbands, served skillfully as the study director for the project. Another senior staff officer, David Hafemeister, provided important technical expertise and collegial support to the effort. Once the report emerged from final classification review, staff officer Christopher Eldridge managed the production of the report, working with the relevant National Academies staff and with program assistant Amy Giamis, who prepared the manuscript. Their collective efforts are much appreciated.

As already mentioned, the issues surrounding the CTBT have a long history and a voluminous literature. In carrying out its study, the committee benefited greatly from this substantial body of prior work, both from classified and open sources; not all these could be cited in the report. The Committee is profoundly grateful for the assistance of the many scholars and government analysts previously and currently engaged in aspects of CTBT issues. It asked me to recognize especially its fruitful interactions with the other studies undertaken for General Shalikashvili by JASON and by the Defense Threat Reduction Agency.

The CTBT committee was also fortunate to receive help from many parts of the Department of State, the Department of Energy, and the Intelligence Community. Staff members from these agencies—and the directors and staff members from the three DOE nuclear-weapon laboratories—were generous with their time in clarifying technical questions and ensuring that the committee had access to the most up-to-date information. The Lawrence Livermore National Laboratory hosted several of the committee's meetings, providing a gracious and productive working environment that was much appreciated.

Last but not least, I would like to thank the members of the CTBT committee. I believe that their report provides an indispensable input to the wider, on-going discussion of nuclear-weapons testing in general, as well as to the U.S. position on the Comprehensive Nuclear Test Ban Treaty.

Bruce Alberts
President, National Academy of Sciences

---

[1] John M. Shalikashvili, *Findings and Recommendations Concerning the Comprehensive Nuclear Test Ban Treaty* (Department of State, January 2001).

ACKNOWLEDGMENTS

This report has been reviewed in draft form by individuals chosen for their diverse perspectives and technical expertise, in accordance with procedures approved by the NRC's Report Review Committee. The purpose of this independent review is to provide candid and critical comments that will assist the institution in making its published report as sound as possible and to ensure that the report meets institutional standards for objectivity, evidence, and responsiveness to the study charge. The review comments and draft manuscript remain confidential to protect the integrity of the deliberative process. We wish to thank the following individuals for their review of this report: Sidney D. Drell, Stanford University; John R. Filson, U.S. Geological Survey; John S. Foster, TRW Inc.; William Happer, Princeton University; John L. Kammerdiener, Los Alamos National Laboratory (Retired); Steven E. Koonin, California Institute of Technology; Hans M. Mark, University of Texas, Austin; Michael M. May, Stanford University; James B. Schlesinger, MITRE Corporation; Charles H. Townes, University of California, Berkeley; Karl K. Turekian, Yale University; Larry D. Welch, Institute for Defense Analyses; Peter D. Zimmerman, U.S. Senate.

Although the reviewers listed above have provided many constructive comments and suggestions, they were not asked to endorse the conclusions or recommendations, nor did they see the final draft of the report before its release. The review of this report was overseen by Sheila E. Widnall, Massachusetts Institute of Technology, and Gerald P. Dinneen, Honeywell Inc. (Retired). Appointed by the National Research Council, they were responsible for making certain that an independent examination of this report was carried out in accordance with institutional procedures and that all review comments were carefully considered. Responsibility for the final content of this report rests entirely with the authoring committee and the institution.

The Committee on Technical Issues Related to the Ratification of the Comprehensive Nuclear Test Ban Treaty operated under the auspices of the Committee on International Security and Arms Control (CISAC), a standing committee of the National Academy of Sciences. For purposes of administration, CISAC is part of the Policy and Global Affairs Division of the National Research Council.

CONTENTS

Executive Summary — 1
- Confidence in the Nuclear-Weapon Stockpile and in Related Capabilities — 1
- Capabilities for Monitoring Nuclear Testing — 5
- Potential Impact of Foreign Testing on U.S. Security Interests and Concerns — 7

Introduction — 13
- The Comprehensive Nuclear Test Ban Treaty — 14
- U.S. Safeguards — 16

Chapter 1: Stockpile Stewardship Considerations: Safety and Reliability Under a CTBT — 19
- Nuclear Testing: Historical Perspective — 20
- Factors Influencing Safety and Reliability — 22
- Elements of an Effective Stewardship Program — 27
- Maintaining Nuclear Design Capabilities — 32
- Change-Control Discipline — 32
- Priorities in Stockpile Stewardship — 33
- Concluding Remarks — 33

Chapter 2: CTBT Monitoring Capability — 35
- General Aspects of All CTBT Monitoring Technologies — 37
- Monitoring Underground Nuclear Explosions — 39
- Radionuclide Releases From Underground Explosions — 45
- Monitoring Against Underground Evasion Scenarios — 46
- Methods For Improving Seismic Monitoring Capability — 49
- Monitoring Underwater Nuclear Explosions — 51
- Monitoring Nuclear Explosions in the Atmosphere — 52
- Monitoring Nuclear Explosions in Space — 53
- The Role of Confidence-Building Measures and On-Site Inspections — 55
- Research and Development in Support of CTBT Monitoring — 56
- Conclusions on CTBT Monitoring Capability — 57

Chapter 3: Potential Impact of Clandestine Foreign Testing: U.S. Security Interests and Concerns — 61
- Two Reference Cases: No CTBT and the CTBT Strictly Observed — 63
- Evasive Testing Under A CTBT — 67
- Assessment of the Impact on U.S. Security Interests Of Nuclear Weapons Tests of Selected Countries — 70
- Summary of Potential Effects of Clandestine Foreign Testing — 77

Appendix A: Biographical Sketches of Committee Members — 79
Appendix B: List of Committee Meetings and Briefings — 83

Executive Summary

This committee's charge was to review the state of knowledge about the three main technical concerns raised during the Senate debate of October 1999 on advice and consent to ratification of the Comprehensive Nuclear Test Ban Treaty (CTBT), namely:

(1) the capacity of the United States to maintain confidence in the safety and reliability of its nuclear stockpile—and in its nuclear-weapon design and evaluation capability—in the absence of nuclear testing;

(2) the capabilities of the international nuclear-test monitoring system (with and without augmentation by national technical means and by instrumentation in use for scientific purposes, and taking into account the possibilities for decoupling nuclear explosions from surrounding geologic media); and

(3) the additions to their nuclear-weapon capabilities that other countries could achieve through nuclear testing at yield levels that might escape detection—as well as the additions they could achieve without nuclear testing at all—and the potential effect of such additions on the security of the United States.

This unclassified Executive Summary provides a synopsis of findings presented at greater length in the unclassified report that follows. Additional detail and analysis are provided in a classified annex.

Confidence in the Nuclear-Weapon Stockpile and in Related Capabilities

We judge that the United States has the technical capabilities to maintain confidence in the safety and reliability of its existing nuclear-weapon stockpile under the CTBT, provided that adequate resources are made available to the Department of Energy's (DOE) nuclear-weapon complex and are properly focused on this task. The measures that are most important to maintaining and bolstering stockpile confidence are (a) maintaining and bolstering a highly motivated and competent work force in the nuclear-weapon laboratories and production complex, (b) intensifying stockpile surveillance, (c) enhancing manufacturing/remanufacturing capabilities, (d) increasing the performance margins of nuclear-weapon primaries, (e) sustaining the capacity for development and manufacture of the non-nuclear components of nuclear weapons, and (f) practicing "change discipline" in the maintenance and remanufacture of the nuclear subsystem.

(a) Attracting and retaining a high-quality work force in the nuclear-weapon complex will require adequate budgets, other clear signals about future program direction and scope, long-term program commitments to technically challenging assignments, and greater attention to quality-of-work-life issues (including the nature of the burdens imposed by necessary protection of national-security secrets). The lack of requirements for new nuclear-weapon designs and the end of nuclear-explosive tests have eliminated some of the traditional technical opportunities in the nuclear-weapon field, but there are many professional challenges and opportunities in maintaining and developing the nuclear-weapon technology and science base for stockpile stewardship under a CTBT and in preparing for possible future weapon development, and there are increasingly powerful diagnostic, analytical, and computational techniques available that can make working on these challenges exciting and productive. A CTBT, in itself, need not prevent attracting and retaining the needed high-quality work force.

(b) The first line of defense against defects in the stockpile that would adversely affect safety and reliability is an aggressive surveillance program. Accordingly, the Stockpile Stewardship Program (SSP) includes an Enhanced Surveillance activity that involves increased focus on the nuclear components, an increased number of diagnostic procedures applied to the weapons that are randomly withdrawn from the stockpile, and increased technical depth of the inspections. While it is prudent to expect that age-related defects affecting stockpile reliability may occur increasingly as the average age of weapons in the stockpile increases in the years ahead, and that such defects may combine in a nonlinear or otherwise poorly specified manner, nuclear testing is not needed to discover these problems and is not likely to be needed to address them.

(c) Remanufacture to original specifications is the preferred remedy for the age-related defects that materialize in the stockpile. This makes it essential that a capability to remanufacture and assemble the nuclear subsystems for nuclear weapons be maintained in the U.S. production complex, with a capacity consistent with best estimates of component lifetimes, stockpile trends, and allowances for occasional unexpected problems. Current estimates, based on projections of the size of the enduring stockpile, indicate that the technical challenges of ongoing repair and remanufacture can be met at existing production-complex sites, provided that their facilities are brought up to and maintained at modern standards of operation. Establishment of a limited-quantity production capability for certified pits at Los Alamos is a particular necessity, as no other facility for this exists in the United States.

(d) A primary yield that falls below the minimum level needed to drive the secondary to full output is the most likely potential source of serious nuclear-performance degradation. Because primary yield margins in these weapons can be increased by changes that would not require nuclear testing, it is possible to use enhanced margins to provide a degree of insurance against minor aging effects and changes in material or process specifications arising in the refurbishment of the weapons. We urge that this be done.

(e) Based on past experience, it is probable that the majority of aging problems will be found in the non-nuclear components of stockpile weapons. Since the non-nuclear components and subsystems can be fully tested under a CTBT, it is possible to incorporate new technologies in these weapon parts as long as these can be shown not to have any adverse effect on proper functioning of the nuclear subsystem. If technologies involved in the non-nuclear components become prohibitively difficult to support with the passage of time because they are no longer utilized in the private sector, needed replacements can be based on current materials, technologies, and manufacturing processes. This does require, however, the provision of adequate resources to provide not only the needed manu-

(f) It is important that a rigorous, highly disciplined process be instituted for controlling changes in the nuclear components. Such a process must discourage deviations from the original specifications. Before adopting deviations that are judged necessary, they must be analyzed thoroughly for potential performance impacts. In the long term, the process must also protect against performance degradations due to cumulative effects of multiple small changes in materials and/or processes that may be introduced in the course of periodic refurbishment operations. The required change-control process must begin with a thorough documentation of the original design and manufacturing specifications. Any subsequent deviations must be thoroughly documented. The resulting audit trail should make it possible to include consideration of possible cumulative effects in judging the acceptability of any proposal for further change. In order to avoid the introduction of interference effects between nuclear and non-nuclear components, prudence dictates that a similar discipline be practiced in regard to any changes in design or location of non-nuclear components situated in proximity to the nuclear subsystem.

Confidence in the safety and reliability of stockpiled nuclear weapons depended far more on activities in the first five categories just described than on nuclear testing even when numbers and kinds of nuclear tests were essentially unconstrained. (The sixth category did not play a large role in the past, because weapons were generally replaced by new tested designs before cumulative changes could become a concern.) Most U.S. nuclear tests were focused on the development of new designs; the other major roles of testing were exploring weapon physics and investigating weapon effects. The so-called stockpile confidence tests were limited to only one per year and–with two exceptions (involving weapon types retired soon after the tests)–they involved new-production units, so they would better be described as "production verification" tests. Even in the absence of constraints on nuclear testing, no need was ever identified for a program that would periodically subject stockpile weapons to nuclear tests.

Stockpile stewardship by means other than nuclear testing, then, is not a new requirement imposed by the CTBT. It has always been the mainstay of the U.S. approach to maintaining confidence in stockpile safety and reliability. The fact that older nuclear designs are no longer being replaced by newer ones means, however, that the average age of the nuclear subsystems in the stockpile will increase over time beyond previous experience. (The average age will eventually reach a maximum that depends on the rate at which weapons are remanufactured or retired.) This means that the enhanced surveillance activities that are part of the current SSP will become increasingly important. But that would be so whether nuclear testing continued or not. Nuclear testing would not add substantially to the SSP in its task of maintaining confidence in the assessment of the existing stockpile.

An important component of the Stockpile Stewardship Program is the development of a broad spectrum of advanced diagnostic tools in support of the surveillance function. These tools are intended to yield a more complete understanding of weapon performance and potential failure modes for nuclear as well as non-nuclear components and subsystems. This effort represents a continuation of the traditional knowledge-based approach to problem solving in the nuclear-weapon program, albeit at a significantly accelerated rate of progress. The SSP can already point to significant successes in that regard, as seen, for example, in the implementation of numerous new, relatively small-scale, measurement and analysis techniques ranging from new bench-top inspection instruments to larger-scale laboratory facilities (including, e.g., accelerated aging tests, novel applications of diamond-anvil cells and ultrasonic resonance, synchrotron-based spectroscopy and diffraction, and subcritical and hydrodynamic tests). All of these provide additional assurance that defects due to design flaws, manufacturing problems, or aging effects will be detected in time to enable evaluation and corrective action if such is deemed necessary.

While the smaller-scale diagnostic developments will remain key to a robust surveillance function, and therefore require continued emphasis, to date most of the debate over the need for new diagnostic tools has focused on larger-scale, capital-intensive experimental and computational facilities currently under development or being planned for the future. Current programs include the Dual Axis Radiographic Hydro Test (DARHT) facility, the National Ignition Facility (NIF), and the Advanced Simulation and Computing (ASC) program. In the immediate future, because of the enormous scientific and engineering challenges associated with the development and eventual utilization of these tools, they can play an important role in helping the nuclear-weapon laboratories attract and retain essential new technical talent. In the longer term they can also be expected to strengthen the scientific underpinnings of nuclear-weapon technology, and thus offer the potential for enlarging the range of acceptable solutions to any stockpile problems that might be encountered in the future. The initial capabilities achieved in the DARHT and ASC programs have already proven to be of value.

Despite these obvious benefits, the importance of this class of tools to the immediate core functions of maintaining an enduring stockpile should not be overstated. In particular, it would be very unfortunate if confidence in the safety and reliability of the stockpile under a CTBT in the next decade or so were made to appear conditional on the major-tool initiatives having met their specified performance goals. Most importantly, their costs should not be allowed to crowd out expenditures on the core stewardship functions, including the capacity for weapon remanufacture, upon which continued confidence in the enduring stockpile most directly depends.

Although a properly focused SSP is capable, in our judgment, of maintaining the required confidence in the enduring stockpile under a CTBT, we do not believe that it will lead to a capability to certify new nuclear subsystem designs for entry into the stockpile without nuclear testing—unless by accepting a substantial reduction in the confidence in weapon performance associated with certification up until now, or a return to earlier, simpler, single-stage design concepts, such as gun-type weapons. Our belief that the introduction of new weapons into the stockpile will be restricted to nuclear designs possessing a credible test pedigree is not predicated on any conjectures as to the likelihood of DARHT, NIF, ASC or other major facilities achieving their design goals. Thus, we do not share the concern that has been expressed by some that these facilities will undermine the CTBT's important role in buttressing the non-proliferation regime.

In the event that quantity replacements of major components of the nuclear subsystem should become necessary, prudence would indicate the desirability of formal peer reviews. Evaluation of the acceptability of age-related changes relative to original specifications and the cumulative effect of individually small modifications of the nuclear subsystem should also be subject to periodic independent review. Such reviews, involving the three weapon laboratories and external reviewers, as appropriate, would evaluate potential adverse effects on system performance and the possible need for nuclear testing.

Nuclear-weapon design activities are not prohibited under the CTBT, and preserving the capability to develop new designs—in case such are ever needed—is a stated goal of U.S. policy, and is one means by which the knowledge of retiring designers is retained. The use of ever more capable computational tools and more realistic material models to understand the relevant database from past nuclear tests, together with the use of advanced hydrodiagnostic techniques to study stockpile-related issues, is an important part of preserving this design capability. The associated design and evaluation expertise will aid in interpreting and perhaps anticipating foreign activities in nuclear-weapon development. We do not believe that nuclear testing is essential to maintaining these design and evaluation capabilities, even though such testing *would* be essential to certifying the performance of new designs at the level of confidence associated with currently stockpiled weapons.

Some have asserted, in the CTBT debate, that confidence in the enduring stockpile will inevitably degrade over time in the absence of nuclear testing. Certainly, the aging of the stockpile combined with the lengthening interval since nuclear weapons were last exploded will create a growing challenge, over time, to the mechanisms for maintaining confidence in the stockpile. But we see no reason that the capabilities of those mechanisms—surveillance techniques, diagnostics, analytical and computational tools, science-based understanding, remanufacturing capabilities—cannot grow at least as fast as the challenge they must meet. (Indeed, we believe that the growth of these capabilities—except for remanufacturing of some nuclear components—has more than kept pace with the growth of the need for them since the United States stopped testing in 1992, with the result that confidence in the reliability of the stockpile is better justified technically today than it was then.) It seems to us that the argument to the contrary—that is, the argument that improvements in the capabilities that underpin confidence in the absence of nuclear testing will inevitably lose the race with the growing needs from an aging stockpile—underestimates the current capabilities for stockpile stewardship, underestimates the effects of current and likely future rates of progress in improving these capabilities, and overestimates the role that nuclear testing ever played (or would ever be likely to play) in ensuring stockpile reliability.

Capabilities for Monitoring Nuclear Testing

Detection, identification, and attribution of nuclear explosions rest on a combination of methods, some being deployed under the International Monitoring System (IMS) established under the CTBT, some deployed as National Technical Means (NTM), and some relying on other methods of intelligence collection together with openly available data not originally acquired for treaty monitoring. The following conclusions presume that all of the elements of the IMS are deployed and supported at a level that ensures their full capability, functionality, and continuity of operation into the future.

In the absence of special efforts at evasion, nuclear explosions with a yield of 1 kiloton (kt) or more can be detected and identified with high confidence in all environments. Specific capabilities in different environments are as follows:

- Underground explosions can be reliably detected and can be identified as explosions, using IMS data, down to a yield of 0.1 kt (100 tons) in hard rock if conducted anywhere in Europe, Asia, North Africa, and North America. In some locations of interest such as Novaya Zemlya, this capability extends down to 0.01 kt (10 tons) or less. Depending on the medium in which the identified explosion occurs, its actual yield could vary from the hard rock value over a range given by multiplying or dividing by a factor of about 10, corresponding respectively to the extremes represented by a test in deep unconsolidated dry sediments (very poor coupling) and a test in a water-saturated environment (excellent coupling). Positive identification as a nuclear explosion, for testing less than a few kilotons, could require on-site inspection unless there is detectable venting of radionuclides. Attribution would likely be unambiguous.

- Atmospheric explosions can be detected and identified as nuclear, using IMS data, with high confidence above 500 tons on continents in the northern hemisphere and above 1 kt worldwide, and possibly at much lower yields for many sub-regions. While attribution could be difficult based on IMS data alone, evaluation of other information (including that obtained by NTM) could permit an unambiguous determination.
- Underwater explosions in the ocean can be reliably detected and identified as explosions, using IMS data, at yields down to 0.001 kt (1 ton) or even lower. Positive identification as a nuclear explosion could require debris collection. Attribution might be difficult to establish unless additional information was available, as it might be, from NTM.
- Explosions in the upper atmosphere and near space can be detected and identified as nuclear, with suitable instrumentation, with great confidence for yields above about a kiloton to distances up to about 100 million kilometers from Earth. (This capability is based on the assumption that relevant instruments that have been proposed for deployment on the follow-on system for the DSP satellites will in fact be funded and installed.) Such evasion scenarios are costly and technically difficult to implement. If they materialize, attribution will probably have to rely upon NTM, including interpretation of missile-launch activities.

The capabilities to detect and identify nuclear explosions without special efforts at evasion are considerably better than the "one kiloton worldwide" characterization that has often been stated for the IMS. If deemed necessary, these capabilities could be further improved by increasing the number of stations in networks whose data streams are continuously searched for signals.

In the history of discussions of the merits of a CTBT, a number of scenarios have been mentioned under which parties seeking to test clandestinely might be able to evade detection, identification, or attribution. With the exception of the use of underground cavities to decouple explosions from the surrounding geologic media and thereby reduce the seismic signal that is generated, none of these scenarios for evading detection and/or attribution has been explored experimentally. And the only one that would have a good chance of working without prior experimentation is masking a nuclear test with a large chemical explosion nearby in an underground mine. The experimentation needed to explore other approaches to evasion would be highly uncertain of success, costly, and likely in itself to be detected.

Thus, the only evasion scenarios that need to be taken seriously at this time are cavity decoupling and mine masking. In the case of cavity decoupling, the experimental base is very small, and the signal-reduction ("decoupling") factor of 70 that is often mentioned as a general rule has actually only been achieved in one test of very low yield (about 0.4 kt). The practical difficulties of achieving a high decoupling factor—size and depth of the needed cavity and probability of significant venting—increase sharply with increasing yield. And evaders must reckon with the high sensitivity of the global IMS, with the possibility of detection by regional seismic networks operated for scientific purposes, and with the chance that a higher-than-expected yield will lead to detection because their cavity was sized for a smaller one.

As for mine masking, chemical explosions in mines are typically ripple-fired and thus relatively inefficient at generating seismic signals compared to single explosions of the same total yield. For a nuclear explosion that is not also cavity-decoupled to be hidden by a mine explosion of this type, the nuclear yield could not exceed about 10 percent of the aggregate yield of the chemical explosion. A very high yield, single-fired chemical explosion could mask a nuclear explosion with yield more comparable to the chemical one, but the very rarity of chemical explosions of this nature would draw suspicion to the event. Masking a nuclear yield even as large as a kiloton in a mine would require combining the cavity-decoupling and mine-masking scenarios, adding to the difficulties of cavity decoupling already mentioned.

Taking all factors into account and assuming a fully functional IMS, we judge that an underground nuclear explosion cannot be confidently hidden if its yield is larger than 1 or 2 kt.

Evasion scenarios have been suggested that involve the conduct of nuclear tests in the atmosphere or at the ocean surface where the event would be detected and identified but attribution might be difficult. NTM of the United States and other nations might provide attribution, without being predictable by the evader.

The task of monitoring is eased (and the difficulty of cheating magnified), finally, by the circumstance that most of the purposes of nuclear testing—and particularly exploring nuclear-weapon physics or developing new weapons—would require not one test but many. (An exception would be the situation in which an aspiring nuclear weapon state had been provided the blueprints for a weapon by a country with greater nuclear weapon capabilities, and might need only a single test to confirm that it had successfully followed the blueprints.) Having to conduct multiple tests greatly increases the chance of detection by any and all of the measures in use, from the IMS, to national technical means, to sensors in use for other purposes.

It can be expected, in future decades, that monitoring capabilities will significantly improve beyond those described here, as instrumentation, communications, and methods of analysis improve, as data archives expand and experience increases, and as the limited regions associated with serious evasion scenarios become the subject of close attention and better understanding. Of course, the realization of this expectation depends on continued U.S. public and policy maker recognition of the importance of this country's capacity to monitor nuclear testing, with concomitant commitments of resources to the task.

Potential Impact of Foreign Testing on U.S. Security Interests and Concerns

The potential impact on U.S. security interests and concerns of the low-yield foreign nuclear tests that could plausibly occur without detection in a CTBT regime can only be meaningfully assessed by comparison with two alternative situations—the situation in the absence of a CTBT, and the situation in which a CTBT is being strictly observed by all parties. The key questions are: How much of the benefit of a strictly observed CTBT is lost if some countries test clandestinely within the limits imposed by the capabilities of the monitoring system? In what respects is the case of limited clandestine testing under a CTBT better for U.S. security—and in what respects worse—than the case of having no CTBT at all? If some nations do not adhere to a CTBT and test openly, how do the technical and political impacts differ from a no-CTBT era?

In these comparisons, two kinds of effects of nuclear testing by others on U.S. security interests and concerns need to be recognized: the *direct* effects on the actual nuclear-weapon capabilities and deployments of the nations that test, with implications for military balances, U.S. freedom of action, and the possibilities of nuclear-weapon use; and the *indirect* effects of nuclear testing by some states on the aspirations and decisions of other states about acquiring and deploying nuclear weapons, or about acquiring and deploying non-nuclear forces intended to offset the nuclear weapons of others. A CTBT, to the extent that it is observed, brings security benefits for the United States in both categories—limitations on the nuclear-weapon capabilities that others can achieve, and elimination of the inducement of states to react to the testing of others with testing and/or deployments of their own.

In the reference case of no CTBT at all, the Nuclear-Weapon States Party to the Non-Proliferation Treaty (NPT) would be able to test without legal constraint in the underground environment (except for the 150-kt limit agreed to by the United States and Russia under the bilateral Threshold Test Ban Treaty), and non-parties to the NPT would similarly be able to test without constraint. Non-Nuclear-Weapon-States Party to the NPT would be constrained legally from testing. In this circumstance:

- China and Russia might use the option of testing to make certain refinements in their nuclear arsenals. In the case of Russia, it is difficult to envision how such refinements could significantly increase the threats to U.S. security interests that Russia can pose with the previously tested nuclear-weapon types it already possesses.
- In the case of China, further nuclear testing might enable reductions in the size and weight of its nuclear warheads as well as improved yield-to-weight ratios. Such improvements would make it easier for China to expand and add multiple independently targetable re-entry vehicles (MIRV) to its strategic arsenal if it wanted to do so, and changes in these directions would affect U.S. security interests. But China could also achieve some kinds of improvements in its nuclear weapons without nuclear testing, and if it wanted to do so it could achieve considerable expansion and MIRVing of its arsenal using nuclear-weapon types it has already tested.
- India and Pakistan could use their option of testing, as non-parties to the Non-Proliferation Treaty, to perfect boosted fission weapons and thermonuclear weapons, greatly increasing the destructive power available from a given quantity of fissile material and the destructive power deliverable by a given force of aircraft or missiles. (Of course they might also do this under a CTBT that they had not signed, but the absence of a CTBT and the resumption of testing by others would make it politically much easier for them to do so.) The likelihood that either of these countries would use nuclear weapons against the United States seems very low, but the United States and its allies would nonetheless have serious concerns about the increase in nuclear-weapons dangers and arms-race potential in and around South Asia that such developments would portend.
- Plausibly larger than the direct effects of testing by Nuclear-Weapon States and non-parties to the NPT in the absence of a CTBT is the potential indirect effect of such testing in the form of a breakdown of the NPT regime, manifested in more widespread testing (by such countries as North Korea, Iraq, and Iran, for example), which could lead in turn to nuclear weapons acquisition by Japan, South Korea, and many others.

A future no-CTBT world, then, could be a more dangerous world than today's, for the United States and for others. In particular, the directions from which nuclear attack on the United States and its allies would have become conceivable—and the means by which such attack might be carried out (meaning not only intercontinental ballistic missiles (ICBM) but also, among others, ship-based cruise missiles, civilian as well as military aircraft, and truck bombs following smuggling of the weapons across U.S. borders)—would have multiplied alarmingly.

In our second reference case of a CTBT scrupulously observed, nuclear threats to the United States could still evolve and grow, but the range of possibilities would be considerably constrained. Boosted fission weapons and thermonuclear weapons would be confined to the few countries that already possess them and to those to which such weapons might be transferred, or to which designs might be communicated with sufficient precision that a trusting and competent recipient might be able to reproduce them. Other countries might have less stringent confidence requirements than does the United States, but, in general, they also are much more limited in the technology available for pursuing an exact reproduction; substitution of materials or techniques might bring uncertainty or even failure. Perhaps most importantly, in a world in which nuclear testing had been renounced and the NPT remained intact, nuclear proliferation would be opposed by a powerful political norm in which Nuclear-Weapon States and other parties to the NPT and CTBT would find their interests aligned.

In the case we now wish to compare to the no-CTBT and rigorously-observed-CTBT reference cases—that of clandestine testing under a CTBT, within the limits imposed by the monitoring system—we distinguish between two classes of potential cheaters, those with greater prior nuclear testing experience and/or design sophistication and those with lesser prior testing experience and/or sophistication. The purposes and plausible achievements for testing at various yields by countries with lesser versus greater prior nuclear test experience and/or design sophistication are summarized in the following table. Table ES-1 describes what could be done, not necessarily what will be done.

Table ES-1 Purposes and Plausible Achievements for Testing at Various Yields

Yield	Countries of lesser prior nuclear test experience and/or design sophistication[*]	Countries of greater prior nuclear test experience and/or design sophistication
Subcritical testing only (permissible under a CTBT)	• Equation-of-state studies • High-explosive lens tests for implosion weapons • Development & certification of simple, bulky, relatively inefficient unboosted fission weapons	same as column to left, plus • limited insights relevant to designs for boosted fission weapons
Hydronuclear testing (yield < 0.1 t TNT, likely to remain undetected under a CTBT)	• one-point safety tests (with difficulty)	• one-point safety tests • validation of design for unboosted fission weapon with yield in 10-ton range
Extremely-low-yield testing (0.1 t < yield <10 t, likely to remain undetected under a CTBT)	• one-point safety tests	• validation of design for unboosted fission weapon with yield in 100-ton range • possible overrun range for one-point safety tests
Very-low-yield testing (10 t < yield < 1-2 kt, concealable in some circumstances under a CTBT)	• limited improvement of efficiency & weight of unboosted fission weapons compared to 1st-generation weapons not needing testing • proof tests of compact weapons with yield up to 1-2 kt (with difficulty)	• proof tests of compact weapons with yield up to 1-2 kt • partial development of primaries for thermonuclear weapons
Low-yield testing (1-2 kt < yield < 20 kt, unlikely to be concealable under a CTBT)	• development of low-yield boosted fission weapons • eventual development & full testing of some primaries & low-yield thermonuclear weapons • proof tests of fission weapons with yield up to 20 kt	• development of low-yield boosted fission weapons • development & full testing of some primaries & low-yield thermonuclear weapons • proof tests of fission weapons with yield up to 20 kt
High-yield testing (yield > 20 kt, not concealable under a CTBT)	• eventual development & full testing of boosted fission weapons & thermonuclear weapons	• development & full testing of new configurations of boosted fission weapons & thermonuclear weapons

States with extensive prior test experience are the ones most likely to be able to get away with any substantial degree of clandestine testing, and they are also the ones most able to benefit

---

[*] That is, lacking an adequate combination of nuclear-test data, advanced instrumentation, and sophisticated analytical techniques, and without having received assistance in the form of transfer of the relevant insights.

technically from clandestine testing under the severe constraints that the monitoring system will impose. But the only states in this category that are of possible security concern to the United States are Russia and China. As already noted, the threats these countries can pose to U.S. interests with the types of nuclear weapons they have already tested are large. What they could achieve with the very limited nuclear testing they could plausibly conceal would not add much to this.

If Russia or China were to test clandestinely, within the limits imposed by the monitoring system, because they thought they needed to do so to maintain the safety or reliability of their enduring stockpiles, this would not add to the threat they would have posed to the United States in the circumstance that they were able to maintain the safety and reliability of their stockpiles without testing. Clandestine testing by Russia or China to maintain their confidence in their stockpile—although in violation of the CTBT, threatening to the non-proliferation regime, and not to be condoned—might actually be *less* threatening to the United States than either their losing confidence in the reliability of their weapons and building up the size of their arsenal to compensate, or their openly abrogating a CTBT in order to conduct the testing they thought necessary to maintain or modernize their stockpiles.

U.S. security could reasonably be judged to be threatened by clandestine Russian and Chinese testing for stockpile reliability only if the Russians and Chinese were able to maintain the reliability of their stockpiles by means of this cheating while the United States, scrupulously adhering to the CTBT, was unable to maintain the reliability of its own stockpile. This is precisely what has been hypothesized by some critics of the CTBT, but we judge (Chapter 1) that the United States has the technical capabilities to maintain the reliability of its existing stockpile without testing. If really serious reliability problems that only could be resolved through testing did materialize in the Russian or Chinese arsenal, moreover, it is unlikely that the degree of testing needed to resolve them could be successfully concealed.

In contrast to the cases of Russia or China, where their substantial prior experience with testing makes it at least plausible that they might be able to conceal some substantial degree of testing at yields below the threshold of detection, states with lesser prior test experience and/or design sophistication are much less likely to succeed in concealing significant tests. This is in part because of the importance of test experience in constructing cavities that can achieve seismic decoupling without leaking radioactivity, and in part because considerable weapon-design experience is required to achieve low yields. Countries with lesser prior test experience and/or design sophistication would also lack the sophisticated test-related expertise to extract much value from such very-low-yield tests as they might be able to conceal. They could lay some useful groundwork for a subsequent open test program in the event that they left the CTBT regime or it collapsed, but they would not be able to cross any of the thresholds in nuclear-weapon development that would matter in terms of the threat they could pose to the United States.

In relation to two of the key "comparison" questions posed at the beginning of this section about the implications of potential clandestine testing, we therefore conclude as follows:

- Very little of the benefit of a scrupulously observed CTBT regime would be lost in the case of clandestine testing within the considerable constraints imposed by the available monitoring capabilities. Those countries that are best able to successfully conduct such clandestine testing already possess advanced nuclear weapons of a number of types and could add little, with additional testing, to the threats they already pose or can pose to the United States. Countries of lesser nuclear test experience and design sophistication would be unable to conceal tests in the numbers and yields required to master nuclear

weapons more advanced than the ones they could develop and deploy without any testing at all.
- The worst-case scenario under a no-CTBT regime poses far bigger threats to U.S. security—sophisticated nuclear weapons in the hands of many more adversaries—than the worst-case scenario of clandestine testing in a CTBT regime, within the constraints posed by the monitoring system.

Introduction

At the request of General John Shalikashvili (U.S. Army, ret.), Special Advisor to the President and the Secretary of State for the Comprehensive Test Ban Treaty (CTBT), our committee conducted, during the summer and fall of 2000, a review of the state of knowledge about the main technical concerns that were raised in the Senate debate of October 1999 on advice and consent to the ratification of the CTBT. The principal such concerns were:

(1) the capacity of the United States to maintain confidence in the safety and reliability of its nuclear stockpile—and in its nuclear-weapon design and evaluation capability—in the absence of nuclear testing;

(2) the capabilities of the international nuclear-test monitoring system (with and without augmentation by national technical means and by instrumentation in use for scientific purposes, and taking into account the possibilities for decoupling nuclear explosions from surrounding geologic media); and

(3) the additions to their nuclear-weapon capabilities that other countries could achieve through nuclear testing at yield levels that might escape detection—as well as the additions they could achieve without nuclear testing at all—and the potential effect of such additions on the security of the United States.

Our committee's analysis of these issues is provided here in unclassified form and elaborated in a classified annex to our report. In arriving at these conclusions, under tight constraints of time and resources, we have benefited greatly from access to the large body of information and analysis on these topics developed through the efforts of the relevant U.S. government agencies, national laboratories, and various prior advisory committees. (Briefings provided to the committee are listed in Appendix B.)

An understanding of these largely technical issues is indispensable as the public and its representatives seek to arrive at a conclusion about the overarching policy question of whether the ratification and entry into force of the CTBT is in the national interest of the United States—that is, whether the United States would be better off with this treaty in force than without it. But that overarching policy conclusion cannot be arrived at on the basis of the technical aspects alone. What is ultimately required is a balancing of all of the benefits of the CTBT for the United States against all of its liabilities.

On the negative side, agreement by the United States to a CTBT implies that the nation will forego any redesign of the nuclear subsystems in its nuclear arsenal that could be undertaken with confidence only with nuclear testing. Of course, irrespective of a CTBT, the country might decide that there are no military requirements for new designs that would justify the costs or other impacts of the changes. In any case, whether arising from a CTBT or not, such a constraint on the introduction of new designs would lead to aging of the stockpile, with possible impact on the safety or reliability of U.S. nuclear weapons unless effectively counteracted by programs of surveillance, repair, and remanufacture as needed. A further set of negative impacts that has been asserted concerning a constraint on new designs is the resulting inability to extend certain safety improvements to weapons not now so equipped and the inability to optimize nuclear weapons for additional specialized missions. Potential positives of U.S. commitment to a CTBT, on the other hand, include limiting further development of the nuclear-weapon capabilities of potential adversaries; reducing the competitive incentive for acquisition or improvement of nuclear weapons by such states; supplementing U.S. capabilities for the detection and attribution of nuclear explosions by others; and increasing the global credibility of U.S. non-proliferation policy and the strength of the non-proliferation regime. Weighing these pros and cons obviously entails political judgments as well as technical ones.

This committee was not asked to address the full range of issues—political as well as technical—germane to a "net assessment" of whether the United States would be better off with a CTBT in force than without it; and we have not tried to do so. But we believe that our analysis of some of the key technical issues in such a determination, as enumerated above, will be helpful to those who must make it. In what follows, we first summarize briefly what the CTBT says and requires, before turning to the three specific issues in our charge—U.S. nuclear-weapon capabilities under a CTBT, the verifiability of the treaty, and effects of a CTBT on the capabilities of countries of potential concern.

The Comprehensive Nuclear Test Ban Treaty

The Comprehensive Nuclear Test Ban Treaty obligates all parties not to conduct "any nuclear weapon test explosion or any other nuclear explosion" in any environment (i.e., in the atmosphere, underwater, underground, or in space) and to refrain from encouraging or helping any other state to carry out such explosions (Article I). The treaty primarily affects the five treaty-designated Nuclear-Weapon States (China, France, Russia, the United Kingdom, and the United States) since the 180 Non-Nuclear-Weapon States Party to the Nuclear Non-Proliferation Treaty (NPT) are already prohibited from nuclear testing (and from developing and producing nuclear weapons, even if without testing) by the NPT. The four states that are not parties to the NPT (Cuba, India, Israel, and Pakistan) would not be able to test if they signed the CTBT but so far only Israel has signed.

Although "nuclear explosion" is not defined in the treaty, Senate testimony by the Secretary of State, the Under Secretary, and the U.S. CTBT negotiator states explicitly that the negotiating record makes clear that all explosions with nuclear yields greater than zero are prohibited.[1] The Nuclear-Weapon States agreed that, while tests with nuclear yields greater than zero (including "hydronuclear tests") were prohibited, weapons-related experiments producing no nuclear yield, including subcritical experiments involving fissile materials and hydrodynamic tests of weapon assemblies without fissile materials, were not banned by the treaty. Similarly, it was understood that the treaty does not cover nuclear reactors, which obtain energy from controlled fis-

---

[1] Testimony by Secretary of State Madeline Albright, Undersecretary John D. Holum, and Ambassador Setphen G. Ledogar, U.S. Senate. 1999. Committee on Foreign Relations. *Final Review of the Comprehensive Nuclear Test Ban Treaty (Treaty Doc. 105-28)*. 106th Congress, 1st session. October 7.

sion reactions, or inertial confinement fusion experiments directed at obtaining energy from nuclear fusion. The prohibition on "any other nuclear explosions" bans testing or use of nuclear explosions for peaceful purposes.

The treaty does not prohibit research or development of nuclear weapons provided no tests are conducted involving a nuclear yield. The non-binding preamble, however, recognizes that the treaty "by constraining the development and qualitative improvement of nuclear weapons and ending the development of advanced new types of nuclear weapons, constitutes an effective measure of nuclear disarmament and non-proliferation in all its aspects."

The treaty is of unlimited duration (Article IX). Each party, however, has the right to withdraw on six months notice if it decides "extraordinary events relating to the subject matter of this treaty have jeopardized its supreme interests."

The treaty will enter into force 180 days after it has been ratified by all of the 44 states listed in the treaty as IAEA-identified possessors of either nuclear weapons or nuclear reactors (Article XIV). The articles to the treaty cannot be subject to reservations by the parties (Article XV). As of July 1, 2002, 41 of the 44 required states have signed the treaty (holdouts are India, North Korea, and Pakistan) and 31 of the 44 have ratified the treaty, including France, Russia, and the United Kingdom.

After the treaty enters into force, it can be amended at a special Amendment Conference by a simple majority of the parties but any party can veto the amendment (Article VII). If accepted, the amendment will enter into force for all parties after ratification by all those voting for it. (A simpler process applies to changes of an "administrative or technical" nature to the protocols and annexes, provided there is no objection.) Every ten years after entry into force, the treaty may be reviewed at a conference of all parties taking into account any new scientific developments relevant to the treaty (Article VIII). A special provision permits any party to request after ten years that the Review Conference consider the possibility of permitting the conduct of underground tests for peaceful purposes. If the Conference decides without objection that such explosions may be permitted, appropriate amendments that will preclude any military benefits from such explosions will be dealt with by the formal amendment process. These provisions make it extremely unlikely that amendments permitting tests for peaceful uses or any other substantive changes will be approved.

The treaty establishes the Comprehensive Nuclear Test Ban Treaty Organization (CTBTO) to ensure implementation of its provisions, including those for international verification of compliance (Article III). The CTBTO—located in Vienna, Austria—consists of a Conference of all parties to the treaty, an Executive Council with 51 members, and a Technical Secretariat. The Conference, which is the formal governing body of the CTBTO, handles treaty-related policy issues and oversees the treaty's implementation by the Executive Council and the Technical Secretariat. The Conference will meet annually and in special session at the request of the Executive Council or a majority of the parties. The Executive Council serves as the regular day-to-day decision-making body responsible for implementing the provision of and compliance with the treaty. The 51 members are elected by the Conference in a manner to assure regional representation and by criteria that will always assure U.S. representation. The Technical Secretariat is responsible for implementing the treaty's verification activities. The CTBTO is already organized on a provisional basis but cannot be fully operational until the treaty enters into force.

The treaty establishes an extensive verification system (Article IV), which is elaborated on in great detail in a protocol. The foundation of the system is the International Monitoring System (IMS) consisting of a global network of seismic, radiological, infrasound, and hydroacoustic sensors that report their data to the International Data Centre (IDC). After initial screening, the IDC makes all of the raw and processed data available to all parties for their own evaluation using in addition whatever information they may have from their own intelligence.

If suspicious events cannot be quickly resolved through consultations and clarification, any party can request an on-site inspection, based on information from the IMS and/or national technical means, such as satellites, in a manner consistent with international law, which would exclude espionage. The request would have to include a specific area not to exceed 1,000 square kilometers. Approval of the on-site inspection requires the positive vote of 30 of the 51 members of the Executive Council. The specific elements of the very intrusive on-site inspection procedures as well as the rights of both the inspectors and the state being inspected are set forth in great detail. If the Executive Council concludes on the basis of evidence before it, that a nuclear explosion has occurred (Article IV), the Conference (or "if the case is urgent" the Executive Council) may bring the issue, including relevant information and conclusions, to the attention of the United Nations (Article V).

As of July 1, 2002, 165 states had signed and 93 states had ratified the CTBT. The United States is a signatory but has not ratified. Under the Vienna Convention on the Law of Treaties, a state that signs a treaty is bound not to "defeat the object and purpose of the treaty"—in this case the prohibition of nuclear explosions—until such time as the state formally announces that it does not intend to ratify the treaty. Although the United States has never ratified the Vienna Convention, it has always accepted the substantive provisions of the Convention as reflecting international law binding on all states. When the Senate failed to give the necessary two-thirds vote of advice and consent to presidential ratification, President Clinton made clear that he would continue to seek approval. The treaty was returned to the Senate Foreign Relations Committee, where it will remain until it is brought up for another vote or is returned to the Executive by a majority vote of the Senate.

Secretary of State Powell indicated in his confirmation hearing in January 2001 that the Bush Administration would not ask the Senate to approve ratification of the CTBT in the current Congressional session, noting that the issue would be examined in the context of the Administration's overall strategic review and that President Bush has indicated that he has no intention of resuming testing since "we do not see any need for such testing in the foreseeable future."[2] While there has been some dispute as to the constitutional status of the issue in the past, it is now generally accepted that a President can announce without any further Congressional action that he will not ratify a treaty and is therefore no longer bound by the Vienna Convention rules. In that case, Congressional recourse would be limited to non-binding resolutions opposing the action and exercise of the power of the purse by refusing to fund testing activities.

U.S. Safeguards

When President Clinton announced U.S. support for a "zero yield" CTBT on August 11, 1995, he established the following specific safeguards that were included in his formal transmittal of the treaty on September 22, 1997 to the Senate for its advice and consent to ratification:

A) The conduct of a Science Based Stockpile Stewardship program to ensure a high level of confidence in the safety and reliability of nuclear weapons in the active stockpile, including the conduct of a broad range of effective and continuing experimental programs.

---

[2] Statement by Secretary of State Designate Colin Powell, U.S. Senate. 2001. Committee on Foreign Relations. *Nomination of Colin L. Powell to be Secretary of* State. 107th Congress, 1st session. January 17.

B) The maintenance of modern nuclear laboratory facilities and programs in theoretical and exploratory nuclear technology that will attract, retain, and ensure the continued application of our human scientific resources to those programs on which continued progress in nuclear technology depends.

C) The maintenance of the basic capability to resume nuclear test activities prohibited by the CTBT should the United States cease to be bound to adhere to this Treaty.

D) The continuation of a comprehensive research and development program to improve our treaty monitoring capabilities and operations.

E) The continuing development of a broad range of intelligence gathering and analytical capabilities and operations to ensure accurate and comprehensive information on worldwide nuclear arsenals, nuclear weapons development programs, and related nuclear programs.

F) The understanding that if the President of the United States is informed by the Secretary of Defense and the Secretary of Energy (DOE)—advised by the Nuclear Weapons Council, the Directors of DOE's nuclear weapons laboratories, and the Commander of the U.S. Strategic Command—that a high level of confidence in the safety or reliability of a nuclear weapon type that the two Secretaries consider to be critical to our nuclear deterrent could no longer be certified, the President, in consultation with the Congress, would be prepared to withdraw from the CTBT under the standard "supreme national interests" clause in order to conduct whatever testing might be required.[3]

---

[3] White House, Office of the President, September 22, 1997.

1

Stockpile Stewardship Considerations: Safety and Reliability Under a CTBT

The United States conducted its last underground nuclear-explosive test in September of 1992. In October of that year President Bush signed into law an amendment to the FY1993 Energy and Water Development Appropriations Act, which placed strict limits on the number and purposes of U.S. nuclear tests and established an initial 9-month moratorium on nuclear testing.[1] President Clinton subsequently extended the moratorium, first through September 1994 and then through the completion of the CTBT negotiations in September 1996. As indicated in the Introduction, the U.S. signature on the CTBT will continue to preclude U.S. testing unless and until the President announces formally that the country does not intend to ratify the treaty, and the position of the new Bush administration thus far has been that no need to resume testing is evident.

These decisions have been motivated by non-proliferation interests. They have been facilitated by the absence of military requirements for new nuclear designs that would require nuclear testing. As a result, the U.S. nuclear-weapon program has undergone a fundamental change from its earlier focus on new designs to its current focus on maintaining an enduring stockpile. The active component of this stockpile comprises eight different nuclear-explosive device designs (i.e., a nuclear subsystem comprising a sealed-pit primary, a secondary, and associated parts inside the weapon case) some of which occur in multiple weaponized versions.

The Department of Energy (DOE) has responded to these developments by restructuring the weapon program into a formal Stockpile Stewardship Program (SSP). This program is intended to ensure the continued safety, reliability, and operational readiness of the enduring stockpile, without nuclear testing, for as long as national policy dictates a need for such weapons.[2] The SSP places increased emphasis on strengthening the scientific understanding of nuclear device performance as well as the aging behavior of weapon materials and components. These efforts are supported by sizable capital-intensive facility investments, including development of high-performance computational simulation and modeling capabilities under the "Advanced Simulation and Computing (ASC)" program.[3] The SSP also addresses the need for enhanced surveillance of stockpile health and for essential manufacturing capabilities.

---

[1] *Energy and Water Development Appropriations Act, 1993*, Public Law 377, 102nd Congress, 2nd session (October 2, 1993), section 507.

[2] U.S. Department of Energy, National Security Administration, *Stockpile Stewardship Plan: Executive Overview* (DOE/NNSA/DP-0141, June 12, 2000).

[3] ASC is the new name for Accelerated Strategic Computing Iniative; see http://www.nnsa.doe.gov/asc/.

A typical modern U.S. nuclear weapon consists of approximately 6,000 individual components. Of these, only the so-called nuclear subsystem, which includes the weapon's "primary" and "secondary," would be subject to the restrictions of a CTBT. The remaining components of a nuclear weapon and its delivery system could be tested at will under a CTBT.

It is assumed explicitly that the enduring stockpile will not depend on the introduction of new nuclear subsystem designs that would require nuclear testing for performance validation. However, the SSP is required to maintain the capability for executing new designs should a compelling need ever arise for such. In the shorter term it is conceivable that nuclear subsystems of established nuclear-test pedigree might be incorporated into new weapons to maintain compatibility with evolving strategic and tactical military delivery systems. This should not introduce uncertainties in weapon performance provided that none of the modifications intrude on the nuclear subsystem beyond established design practices.

The expectation that the United States will ratify the CTBT at some point in the future has fueled concerns in some quarters that confidence in the stockpile will inevitably erode over time in the absence of nuclear testing, notwithstanding the SSP. The fact that the stockpile has declined both in total numbers as well as in numbers of weapon types has added to this concern. In the remainder of this chapter we address these issues in four steps: a historical perspective on nuclear testing; a discussion of the factors affecting nuclear-weapon safety and reliability in the context of a CTBT; an analysis of five elements of an effective stockpile stewardship program; and a treatment of some issues in priority-setting in stockpile stewardship. We do not offer here a comprehensive or detailed assessment of the initiatives and facilities of the SSP—this would have been beyond our mandate—but confine ourselves mainly to the effects, on stewardship requirements and effectiveness, of a cessation of nuclear-explosive tests.

Nuclear Testing: Historical Perspective

Since 1945 the United States has amassed an extensive data, knowledge, and experience base derived from over 1,000 individual test explosions. The first 20 years were marked by rapid progress as measured by increasing yields, increasing yield-to-weight ratios, and advances in other militarily significant features. During the 1960s and 1970s tailored output concepts were explored, such as reduced residual-radiation, enhanced neutron, and hot X-ray devices. None of these advanced concepts, however, attracted lasting support from the military services, and none is part of today's enduring stockpile. During the 1970s and early 1980s the advances in nuclear-explosive technology reached a plateau as the nuclear designs of interest to the services approached performance limits set by the laws of physics. It is during this same period that the current stockpile designs evolved.

From a knowledge standpoint, nuclear-test data obtained during the past 25 years are of greatest relevance since many of these bear directly on the designs in the enduring stockpile and were obtained from well-diagnosed tests. Most nuclear tests were focused on the development of new designs—ranging from exploratory concepts to designs aimed at specific requirements promulgated by the military—although only a small fraction of the warhead designs subjected to nuclear testing were identical to the warhead designs that actually entered the stockpile. Other tests centered on weapon-physics issues not related to a specific weapon-development program. Some pursued novel ideas, such as the X-ray laser. Many primaries utilized in nuclear tests conducted for a variety of these purposes, however, were taken from the stockpile production line, which enabled these tests to contribute something to stockpile confidence. Certainly, the totality of the nuclear testing experience contributed to design and manufacturing expertise and to an overall sense of assurance that the U.S. design and production complex was delivering reliable weapons. But the number of tests was too small to provide a statistical basis for confidence, and it did not allow coverage, for each design, of the range of stockpile-to-target sequence (STS) conditions called for in the military specifications.[4]

A test of one stockpile system per year became institutionalized in the last decade of U.S. nuclear testing. These so-called stockpile confidence tests involved new-production units, with two exceptions (involving weapon types that were low on the priority lists and were retired soon after the tests). Since these tests primarily established confidence in the stockpile system as it was initially produced, validating the integrated performance of the nuclear subsystem at stockpile entry, they should be described more correctly as "production verification" tests. Any deviations from the expected results were attributed to final design and/or fabrication variations from units tested during development, certainly not to aging effects. Even in the absence of constraints on nuclear testing, no need was ever identified for a program that would periodically subject stockpile weapons to nuclear tests.

Thus, nuclear testing never provided—and was never intended to provide—a statistical basis for confidence in the performance of stockpiled weapons. Rather, it has always been deemed adequate to rely on inspection of the nuclear components as part of a systematic stockpile surveillance program, after the nuclear components had initially been tested. This reliance on surveillance is necessary because nuclear tests have always been too few to provide a statistically significant measure to assess weapons reliability quantitatively. Thus, nuclear testing in itself is intrinsically ill-suited to monitor the health of the stockpile. The question of whether nuclear testing might ever be needed to correct problems discovered in weapons after certification and deployment generated some controversy during the 1980s.[5] However, it was shown that almost all of the problems cited in support of this proposition were either of a kind not requiring nuclear testing to correct or represented cases where testing had been inadequate during development.[6] In relating these experiences to the current situation it is also important to note that the observed failures all occurred within three years after entry into the stockpile. The weapons in today's active stockpile have long passed the age where anomalies in initial production units are a significant problem. Furthermore, they are all based on tested designs that have taken advantage of lessons learned from older vintages. Having analyzed the historical record, we judge that aging problems that may occur in the future will be detected in DOE's surveillance program and that they can be corrected by replacement of the affected components, given adequate production facilities.

It should be recognized that the production run for a stockpiled system includes a range of variations—deviations from the ideal. The refurbishment and/or remanufacture of nuclear-subsystem components will involve similar variations. The effects of such variations in both cases can be expected to be small, although they are not reliably calculable. This is because the systems in the U.S. enduring stockpile are robust, meaning not very sensitive to small variations from design specifications or conditions. Straightforward practical measures, moreover, which

---

[4] See, e.g., J. S. Foster (Chairman), et al., *FY1999 Report of the Panel to Assess the Reliability, Safety, and Security of the United States Nuclear Stockpile* (Washington DC, November 8, 1999), p 2.

[5] J. W. Rosengren, *Some Little-Publicized Difficulties With a Nuclear Freeze* (Report Prepared by R&D Associates for the Office of International Security Affairs, U.S. Department of Energy, October 1983), and *Stockpile Reliability and Nuclear Test Bans: A Reply to a Critic's Comments* (RDA-TR-138522-001, R&D Associates, November 1986).

[6] R. E. Kidder, *Evaluation of the 1983 Rosengren Report from the Standpoint of a Comprehensive Test Ban (CTBT)*, UCID-20804, Lawrence Livermore National Laboratory, June 1986, and *Stockpile Reliability and Nuclear Test Bans* (UCID-20990, Lawrence Livermore Laboratory, February 1987).

have been validated by previous nuclear tests, can enhance primary performance margins, compensating for the cumulative effect of many small variations due to aging or remanufacture. These measures are discussed in the section on "Performance Margins of the Primary."

Critics of a CTBT have pointed to examples of the risk of introducing new designs into the stockpile without testing. We do not believe that this should be done; we accept that a constraint on U.S. freedom of action imposed by a CTBT would be to preclude the certification of new designs for addition to the stockpile with confidence in performance equal to that associated with the weapons now there. But previously tested nuclear subsystems or components of these could be used in modified or entirely new delivery systems, provided that any influences of the altered system on the nuclear subsystem were deemed to be within acceptable limits in the light of prior nuclear-test experience. The key constraint on deployment of additional nuclear weapons of previously tested designs under a CTBT, should the country wish to do this, would not be the absence of the ability to conduct new nuclear tests but the lack of capacity for manufacturing new primary pits. (The exception to this constraint is one previously tested design that is based on reuse of primary pits from a retired system.)

Critics also assert the difficulty of maintaining weapon-laboratory competence in the absence of nuclear testing. We believe that the absence of new programs of nuclear-weapon development for deployment is a bigger part of the challenge of recruitment and retention than is the absence of testing *per se*. Whether nuclear weapons are seen as an important national asset is also important, as are working conditions at the weapons laboratories. How this challenge can be addressed is discussed in the section on "Human Talent Pool."

The fact that older nuclear designs are no longer being replaced by newer ones means that the average age of the nuclear subsystems in the stockpile will increase over time beyond previous experience. (The average age will eventually reach a maximum that depends on the rate at which weapons are remanufactured or retired.) This means that the enhanced surveillance activities that are part of the current SSP will become increasingly important. But that would be so whether nuclear testing continued or not. Nuclear testing would not add substantially to the SSP in its task of maintaining confidence in the assessment of the existing stockpile.

Finally, we note that it has never been possible for practical reasons to explore the full operating range specified by the "military characteristics" during the development phase of nuclear designs that entered the stockpile. Computational modeling, using codes and computers that were often primitive by modern standards, was relied on to interpolate and extrapolate the test data over the specified range. In this manner estimates were generated for primary yield variations with boost-gas age, secondary yield variations, and other critical performance measures covering the extremes of the stockpile-to-target-sequence requirements. Modern simulation capabilities have significantly improved the ability to predict the performance of nuclear weapons.

Factors Influencing Safety and Reliability

To understand more fully why confidence in the stockpile historically has remained very high even though nuclear confidence testing has not played a significant role in substantiating this confidence, it is necessary to understand the factors that enter into the assessment of weapon safety and reliability. Nuclear weapons with their thousands of individual components are complex both in function and design. Failure of a critical component can lead to total system failure. This has led to assertions that, in the absence of nuclear testing, the safety and reliability of the weapon cannot be guaranteed. This belief is erroneous on several counts. Most importantly, the vast majority of weapon components are not integral to the nuclear subsystem and can therefore be tested under a CTBT. Furthermore, redundancies in system design guard against the most probable failure modes and, as regards safety, designs of current weapon safing, arming, fuzing, and firing systems are based on fail-safe architectures as described in the following section.

Safety

Safety criteria for nuclear weapons fall into two categories: nuclear-detonation safety and plutonium-dispersal safety. While accidental dispersal of plutonium poses a far less catastrophic threat than does a nuclear detonation, the cleanup costs for a wide-area plutonium dispersal would still be substantial. Over the years, solutions have been incorporated in weapon designs that have reduced the risk of an accidental nuclear detonation to a very low level. Progress has also been achieved regarding plutonium dispersal. It remains true, nonetheless, that operational and logistical measures that minimize weapon exposure to serious accidents still form the most important line of defense, one that is both cost-effective and meets applicable standards if rigorously implemented.

Operational and logistical safety standards have been established for two distinct environments. **Normal environments** are defined in the stockpile-to-target sequence and military-characteristics specifications as those in which the weapon is required to survive without degradation in reliability. **Abnormal environments** are those in which the weapon is not expected to retain its reliability, as would be the case for accidents in which a weapon is subjected to severe damage.

- For **normal-environment** stockpile-to-target sequence scenarios both nuclear-detonation safety as well as plutonium-dispersal safety are determined entirely by the characteristics of the electrical subsystem with its various electrical and mechanical safing, arming, fuzing, and firing components. To prevent inadvertent firing of the primary detonators the weapon electrical subsystem incorporates a degree of fault tolerance in its design that is believed to satisfy the requirement of no greater than a one in one billion (10^{-9}) probability over a warhead's lifetime.

- For **abnormal-environment** stockpile-to-target sequence scenarios one-point nuclear-detonation safety is the only safety criterion that hinges on the nuclear performance of the nuclear subsystem. (This criterion specifies that the probability of a nuclear yield greater than 4 pounds TNT equivalent must not exceed 10^{-6} in the event that a detonation is initiated at any point in the high explosive surrounding the primary.) Confidence that the one-point criterion is satisfied derives from the nuclear-test-based pedigree of primaries in the existing stockpile, all of which have been certified by calculation and testing to be one-point safe. It follows that even under abnormal conditions nuclear-detonation safety is dominated by the electrical subsystem architecture. Analysis of accidents involving nuclear weapons that occurred during the 1950s and 1960s raised serious concerns that the electrical subsystems of that era could conceivably cause a nuclear detonation under the influence of fire, impact, or other accident-related assaults. While these concerns could not be quantified, it became clear that the safety margins under a wide range of credible abnormal conditions fell far short of the 10^{-6} per incident requirement established in 1958. This led initially to operational changes that reduced the probability of accidents. It was followed in the early 1970s by the development of an "Enhanced Nuclear Detonation Safety" (ENDS) electrical-subsystem architecture with predictable fail-safe characteristics.[7] Although the fundamental weak-link/strong-link concept for isolating sources of electrical energy from the primary detonators has remained unchanged, the design and location of the safety-critical links in this architecture have evolved over time. In some cases links have been integrated into the detonator design, thus placing them as close to the high explosive as is functionally possible. While not directly testable for all conceivable accident scenarios, confidence in the robustness of the architecture and its various implementations is firmly rooted in experiment and analysis. Currently a large majority of U.S. weapons feature ENDS subsystems. There are no plans to retrofit weapons lacking ENDS because it is anticipated that these will in time be retired from the stockpile.

While the probability of inadvertent nuclear detonations has been reduced by modern electrical-subsystem designs to acceptably low levels there still remains the threat of plutonium dispersal in accidents. The consequences of such dispersal become greatly amplified if the high explosive detonates, causing the plutonium to become aerosolized. Beginning in 1974, the incorporation in nuclear primary designs of "insensitive" high explosives (IHE) that fail to detonate at rigid-target impact velocities as high as 2,000 feet/second has provided an effective solution to that problem. Not every existing weapon utilizes IHE, primarily because of performance considerations arising from the lower energy density of IHE. Decisions to use conventional high explosive instead of IHE were supported by judgments that the formal weapon-system requirements could not be met with an IHE design, and that the existing safety margins would be adequate for the intended application. Even with IHE, however, pit melting in high-temperature fires and subsequent dispersal of plutonium oxidation products remain as a serious, though lesser, concern. Several stockpile weapons, as well as a nuclear-test-validated alternate weapon system, feature so-called "fire-resistant" pit designs that provide containment of molten plutonium in fuel-fire environments. Nonetheless, it is clear that the most effective way to prevent plutonium dispersal is to minimize the likelihood of energetic accidents that would cause plutonium to be spread beyond the accident site. The exposure to such accidents is greatest during weapon transport, and the potential consequences are much greater for air shipments than for road or rail shipments in protective transporters.

We therefore conclude that the safety of the enduring stockpile as designed is not dependent on nuclear testing. With the exception of one-point safety, which is an intrinsic design feature of all U.S. primaries, nuclear detonation safety is entirely a function of electrical subsystem design. The introduction of IHE and "fire-resistant" pits in some weapon types has resulted in significant reduction of the risk of plutonium dispersal. Adding one or both of these features to a conventional primary requires a new primary design and is therefore not possible in the absence of testing unless a previously validated system having these features can be incorporated into a modified delivery system.

Improvements in ENDS implementations as well as in other command-and-control functions such as use-control will undoubtedly evolve over time. This includes the important matter

---

[7] The safety-critical electrical and electro-mechanical components in the ENDS architecture are located within a structurally robust exclusion volume. The power source required to initiate detonation of the primary high explosive is isolated from the weapon detonators by a "strong-link" switch that is only closed when the weapon is intentionally armed during its delivery trajectory. This switch is also designed to remain open prior to arming, under any conceivable accident scenario, for a length of time long compared to the time required for a "weak link" to cause collapse of the power source. A high-voltage capacitor that stores the charge needed to fire the detonators is a simple example of such a weak link. It can be designed to fail predictably under high-temperature and/or energetic-impact conditions as might be experienced, for example, in an aircraft crash. By placing two weak-link/strong-link sets in series one can achieve confidence that the 10^{-6} failure criterion is satisfied. In practice, the two sets employ very different designs in order to eliminate the possibility of common-mode failures.

of preventing unauthorized use. These improvements can be incorporated without nuclear testing as long as they only affect components outside the nuclear subsystem. From time to time ideas for improvements have been advanced that involve placement of components inside the nuclear subsystem. It is conceivable that these could be accommodated without nuclear testing provided the required mass/volume perturbations are judged to be so small as to pose no threat to the performance of the nuclear subsystem. On the other hand, more intrusive concepts, especially those aimed at increasing plutonium-dispersal safety by means of primary configurations that would keep the fissile material separated from the high explosive until the weapon is armed, could not be implemented without nuclear testing and would therefore be unavailable under a CTBT. Considering the high cost involved, serious reliability concerns, and the uncertain benefit to be derived, however, it seems questionable whether incorporation of any of these ideas into the enduring stockpile could be justified even if nuclear testing were permitted.[8]

Reliability

The reliability of the stockpile is derived from reliability models for each weapon type. These models decompose each weapon type into its functional parts, which are assigned reliabilities derived from the available data. A fault-tree methodology is used to estimate the overall system reliability.

Nuclear subsystems in the enduring stockpile historically have been certified to achieve the specified yield range with 100 percent certainty over the entire range of specified stockpile-to-target sequences, provided that the firing, neutron-generator, and boost-gas subsystems function within their specified tolerances corrected for the range of STS conditions. Although the physics governing the detonation of the nuclear subsystem is exceedingly complex, the engineering complexity in terms of parts count and structural features is for the most part not very great. This has permitted an approach to guaranteeing reliable operation in which the manufacture of the nuclear components and their assembly is carefully controlled within tight tolerances, yielding production versions that replicate devices whose pedigree was established through nuclear tests during the development phase. Periodic inspection of nuclear subsystem parts from weapons withdrawn randomly from the stockpile, together with remedial action when required, ensures that conformance to specification is maintained over time. Analysis of nuclear-test diagnostics data for the systems in the U.S. enduring stockpile has shown that none operate near catastrophic failure thresholds. In addition, there are available measures to increase confidence in nuclear subsystem reliability. These measures are discussed in the section on "Performance Margins of the Primary."

It follows that the formally assessed system reliability of existing systems is determined entirely by the "non-nuclear" subsystems. This includes the safing/arming/fuzing/firing subsystem, the neutron-generator subsystem, the boost-gas handling subsystem, and a large variety of other electrical, mechanical, and structural parts. The parts count runs to several thousand and many of the active components are complex both in design and function. The "non-nuclear" parts do not require nuclear testing for full functional evaluation and a CTBT would therefore

---

[8] The statement transmitted to the Armed Services and Appropriations Committees of the House and Senate by President George H. W. Bush on 19 January 1993 upon signing the law initiating a 9-month moratorium on U.S. tests declared that "we do not believe it would currently be cost-effective to incorporate [additional safety improvements] in the existing stockpile." Undersecretary of Defense John Deutch testified to the House Armed Services Committee on 3 May 1993, early in the Clinton administration, to the same effect (Testimony by J. Deutch, U.S. House of Representatives. 1993. Armed Services Committee. Panel on Military Application of Nuclear Energy. 103rd Congress, 1st session. May 3).

not impact weapon-reliability assessments. (Ensuring conformance to radiation-hardening specifications under a CTBT is discussed below.) The determination of failure probabilities for the functional blocks in the weapon reliability model is based on data obtained from random-sample tests. These are conducted on parts withdrawn at various stages during production, on parts obtained from newly assembled warheads (New Material Evaluation), and on parts obtained from warheads withdrawn at varying intervals from the stockpile (Stockpile Evaluation).

In order to meet the stated goal of affirming, at 2-year intervals, a 90 percent level of confidence that if 10 percent or more of the warheads of any given type contained a flaw this would be detected, 21 warheads of each type with a population under 500 and 22 warheads of each type with a population of 500 or more are withdrawn randomly from the stockpile every 2 years for examination. These weapons are disassembled and the parts inspected, including the nuclear components (one per weapons type is destructively disassembled each year). The non-nuclear parts from approximately 70 percent of these weapons are subjected to extensive functional tests. The remaining 30 percent of the disassembled weapons are subsequently converted into "Joint Test Assembly" (JTA) units and flight-tested. A JTA is a weapon in which the nuclear subsystem has been removed and replaced, typically by a diagnostic telemetry system with closely matching mass properties. The recent introduction of "high-fidelity" test vehicles in which the special nuclear materials have been replaced by matching surrogates allows even more realistic tests of integrated system function. Depending on the particular configuration, JTA flight tests of a bomb or missile enable various weapon functions from arming to detonation of the primary explosive to be evaluated for realistic stockpile-to-target sequence delivery scenarios. These and other (e.g., ground-based) system-level tests have proven to be especially important over the years.

Weapon military characteristics generally include requirements for hardening against nuclear radiation (neutron, X-ray, gamma ray) to reduce vulnerabilities to defensive nuclear bursts (e.g., from a nuclear-armed anti-ballistic missile) and to prevent fratricide in the case of dense targeting. Reliability assessments for weapons in such "hostile" environments have in the past benefited from underground nuclear-effects tests. These tests involved exposure of material samples, components, subsystems, and occasionally full-up reentry systems (minus fissile components) to the radiation of a low-yield nuclear device. A large database on weapons effects has resulted from experiments carried out by the Department of Defense and DOE. Although this database remains of great value to this day, laboratory simulators (pulsed nuclear reactors, electron-beam sources, Bremsstrahlung X-ray sources, lasers, light-initiated explosives, etc.) combined with computational modeling had begun to replace underground nuclear-effects tests for radiation hardening purposes even before the moratorium. The reasons were not only related to the lower cost and greater accessibility of laboratory simulators. For some weapon effects the laboratory simulations were shown to have superior fidelity due to the fact that the radiation from underground explosions had characteristics quite unlike the postulated threat. As a result of these efforts the current stockpile is believed to meet the vulnerability requirements established at the height of the Cold War. Laboratory simulation and analysis as new technologies are introduced into future non-nuclear component designs can provide verification of these radiation hardness levels under a CTBT.

The previous discussion allows two important conclusions to be drawn regarding existing weapon reliability: (1) Weapon reliability is dominated by the non-nuclear system elements, which are testable under a CTBT; and (2) Confidence in the continued reliability of the nuclear subsystem can be maintained by ensuring that in any future rebuilds, especially of the primary, the original specifications are adhered to closely. Because of the potential difficulty, but crucial need, for adhering to original specifications, it is important to recognize that additional measures are available to increase confidence in system reliability (see the section on "Performance Margins of the Primary").

Elements of an Effective Stewardship Program

The Secretaries of Energy and Defense are required to submit to the President an annual weapon-stockpile certification report.[9] As in the past, the report for 2000 certified that all safety and reliability requirements are being met without need for nuclear-yield testing.[10] The analysis presented above makes it possible to define five imperatives for the nuclear-weapon complex that will enable it to provide this assurance into the indefinite future under a permanent test ban.

Human Talent Pool

It is self-evident that a highly motivated, competent work force throughout the complex, supported by a modern infrastructure and adequate budgets, is of overarching importance.[11] The three weapon laboratories and the production complex have found it increasingly difficult to retain their top technical talent and to recruit replacements. Reasons for this vulnerability include attractive career opportunities in the private sector, uncertainties about the future of the nuclear-weapon program, and burdensome restrictions such as those imposed following recent security incidents. Clear signals about future program direction and scope, long-term program commitments to technically challenging assignments, and greater attention to quality of work-life issues should assist in reversing this troubling trend.

The challenge of technical staff retention, recruitment, and training is especially pressing in the core weapon areas at the two nuclear design laboratories (Los Alamos and Lawrence Livermore). The lack of requirements for new nuclear-weapon designs and the end of nuclear-explosive tests have eliminated some of the traditional technical opportunities in the nuclear-weapon field. At the same time it is clear that the importance of the nuclear-weapon experience base will not diminish; rather it is likely to grow. As pointed out earlier, even when nuclear testing was permitted, confidence in the reliability of the nuclear subassembly ultimately rested on the informed judgments of a competent technical staff. It would have been impossible, for example, to establish manufacturing tolerances on the basis of nuclear testing alone. In the future the same informed judgment would be needed to decide at what point nuclear components in the aging stockpile need to be replaced. Despite best efforts it is likely to prove impossible in the future to adhere exactly to the original material and process specifications in the remanufacture of nuclear components. Any changes in the specifications will have to be of a nature that does not perturb in a significant way the nuclear-test pedigree of the design; that is, they will need to remain well within the range of parameters where subsystem behavior is understood based on

---

[9] In preparing the certification report, the Secretary of Energy receives assessments from the Directors of the three weapon laboratories and the Secretary of Defense receives assessments from the Strategic Command and the military services.

[10] The reports for 2001 and 2002 have not been completed as of July 2002.

[11] An extensive discussion of recent circumstances relating to this issue is available in the report of the Chiles Commission (H. G. Chiles, et al., *Report of the Commission on Maintaining U.S. Nuclear Weapons Expertise*, Washington DC, 1 March 1999). The specific work force impacts of recent security measures are discussed in the report of the Commission on Science and Security chaired by former Deputy Secretary of Defense John Hamre (Center for Strategic and International Studies, *Science and Security in the 21st Century: A Report to the Secretary of Energy on the Department of Energy Laboratories*, Washington, DC: CSIS, April 2002).

nuclear-test data. Defining the boundaries of permissible change, then, will again require informed judgments.

There is no reason to fear that talent will be unavailable in the near term (several years) to support any decisions that might have to be made concerning aging effects on nuclear components. As the current experience base erodes due to retirements or resignations, however, a gradual transition will by necessity take place to reliance on experts with no prior "hands-on" nuclear-test experience. We believe that this need not pose a fundamental problem, for several reasons. The extensive U.S. nuclear-test database will continue to be accessible and be capable of providing much additional understanding through further analysis. We also note that the existing knowledge base has been validated not only against nuclear-test data, but against laboratory-scale hydrodynamic experiments, computational simulation and modeling, as well as a broad spectrum of supporting scientific and technological investigations. During recent years dramatic advances have occurred in diagnostic and analytical tools and techniques. Today's weapon codes and computers provide capabilities in speed and fidelity far beyond the state of the art as it existed at the time current weapons first entered the stockpile. New hydrodiagnostic capabilities, deep-penetration radiography (core punch), and high-resolution multi-beam velocimetry are part of the current state of the art. In combination with higher-fidelity simulations of production hardware these capabilities can be used to update our experimental understanding. The "science-based" component of DOE's SSP will provide more realistic material-property models and further improvements in weapon codes. These should lead to a deeper, more detailed understanding of the nuclear-test database and thus provide the underpinnings for a rational decision process concerning nuclear-performance issues as they may surface in the future. They would also help in maintaining a capability to produce new designs if ever needed.

An important component of the Stockpile Stewardship Program is the development of a broad spectrum of advanced diagnostic tools in support of the surveillance function. These tools are intended to yield a more complete understanding of weapon performance and potential failure modes for nuclear as well as non-nuclear components and subsystems. This effort represents a continuation of the traditional knowledge-based approach to problem solving in the nuclear-weapon program, albeit at a significantly accelerated rate of progress. The SSP can already point to significant successes in that regard, as seen, for example, in the implementation of numerous new, relatively small-scale, measurement and analysis techniques ranging from new bench-top inspection instruments to larger-scale laboratory facilities (including, e.g., accelerated aging tests, novel applications of diamond-anvil cells and ultrasonic resonance, synchrotron-based spectroscopy and diffraction, and subcritical and hydrodynamic tests). All of these provide additional assurance that defects due to design flaws, manufacturing problems, or aging effects will be detected in time to enable evaluation and corrective action if such is deemed necessary.

While the smaller-scale diagnostic developments will remain key to a robust surveillance function, and therefore require continued emphasis, to date most of the debate over the need for new diagnostic tools has focused on larger-scale, capital-intensive experimental and computational facilities currently under development or being planned for the future. Current programs include the Dual Axis Radiographic Hydro Test (DARHT) facility, the National Ignition Facility (NIF), and the Advanced Simulation and Computing (ASC) program. In the immediate future, because of the enormous scientific and engineering challenges associated with the development and eventual utilization of these tools, they can play an important role in helping the nuclear-weapon laboratories attract and retain essential new technical talent. In the longer term they can also be expected to strengthen the scientific underpinnings of nuclear-weapon technology, and thus offer the potential for enlarging the range of acceptable solutions to any stockpile problems that might be encountered in the future. The initial capabilities achieved in the DARHT and ASC programs have already proven to be of value.

Despite these obvious benefits, the importance of this class of tools to the immediate core functions of maintaining an enduring stockpile should not be overstated. In particular, it would be very unfortunate if confidence in the safety and reliability of the stockpile under a CTBT in the next decade or so were made to appear conditional on the major-tool initiatives having met their specified performance goals. Most importantly, their costs should not be allowed to crowd out expenditures on the core stewardship functions, including the capacity for weapon remanufacture, upon which continued confidence in the enduring stockpile most directly depends.

Although a properly focused SSP is capable, in our judgment, of maintaining the required confidence in the enduring stockpile under a CTBT, we do not believe that it will lead to a capability to certify new nuclear subsystem designs for entry into the stockpile without nuclear testing—unless by accepting a substantial reduction in the confidence in weapon performance associated with certification up until now, or a return to earlier, simpler single-stage design concepts, such as gun-type weapons. Our belief that the introduction of new weapons into the stockpile will be restricted to nuclear designs possessing a credible test pedigree is not predicated on any conjectures as to the likelihood of DARHT, NIF, ASC, or other major facilities achieving their design goals. Thus, we do not share the concern that has been expressed by some that these facilities will undermine the CTBT's important role in buttressing the non-proliferation regime.

Stockpile Surveillance

The first line of defense against defects in the stockpile that would adversely affect safety and reliability is an aggressive surveillance program. Stockpile surveillance has always been a critical part of the weapon program and it is imperative that all of its traditional activities, including JTA flight tests, continue to be fully supported. This need has been recognized and, accordingly, the SSP includes an Enhanced Surveillance activity that involves increased focus on the nuclear components, an increased number of diagnostic procedures applied to the weapons that are randomly withdrawn from the stockpile, and increased technical depth of the inspections.

In the past, changing military requirements usually caused weapons to be retired before aging became a pressing issue. Aside from occasional anomalies in initial production units, significant aging effects were only rarely encountered. In contrast, one may anticipate that weapons in the enduring stockpile will have to be maintained for the indefinite future. It is prudent to expect that age-related defects affecting stockpile reliability may occur increasingly as the average age of weapons in the stockpile increases in the years ahead, and that such defects may combine in a nonlinear or otherwise poorly specified manner, but nuclear testing is not needed to discover these problems and is not likely to be needed to address them. The rigor applied to understanding the nature and cause of such defects will become an increasingly important aspect of lifetime extension activities. The ability to predict age-related problems before they occur in the stockpile will assist in planning and executing necessary remedial actions. Based on past experience it is clear that the majority of aging problems will be found in the non-nuclear components. Nuclear testing will not be needed to address these problems.

The study of aging phenomena will also provide fertile ground for scientifically challenging research, as recent experience has already demonstrated, thereby contributing to staff retention. Much of the relevant science deals with topics that lie outside the classified domain. This opens up opportunities for collaborations with the broader scientific and engineering communities. Vigorous pursuit of these opportunities, including funding of university research that is done in collaboration with laboratory scientists and engineers, will add greatly to the vitality of the enterprise.

Nuclear-Subsystem Remanufacturing Capability

Remanufacture to original specifications is the preferred remedy for the age-related defects that materialize in the stockpile. This makes it essential that a capability to remanufacture and assemble the nuclear subsystems for nuclear weapons be maintained in the U.S. production complex (LANL, Y-12, Pantex, etc.).[12] The production capacity must be consistent with best estimates of component lifetimes based on current knowledge, realistic projections of future stockpile trends, and allowances for occasional unexpected problems. Current estimates, based on projections of the size of the enduring stockpile, indicate that the technical challenges of ongoing repair and remanufacture can be met at existing production-complex sites, provided that their facilities are brought up to and maintained at modern standards of operation. Establishment of a limited-quantity production capability for certified pits at Los Alamos is a particular necessity, as no other facility for this exists in the United States. Major facility expansion is only likely to be needed in the event of unforeseen problems requiring rapid manufacture of large numbers of stockpile weapons or rapid manufacture of large numbers of new ones. For example, the capability for manufacture or remanufacture of large numbers of pits does not currently exist, nor is it planned for the future. Acquiring such a capability would be an arduous, expensive, and lengthy process. Fortunately, the sealed pit is probably the most stable component of U.S. nuclear subsystems. Lifetime estimates for these pits are currently in excess of 50 years. Accelerated-aging tests now underway should lead to more certain estimates within a few years.

In any refurbishments of the nuclear subsystem, changes in materials and processes from the original specifications should be minimized and in all events, as already noted, must remain within the parameter range explored in prior testing. To minimize the need for substitutions it may become necessary to establish in-house sources for materials (e.g., adhesives) that are no longer obtainable from commercial sources. Materials that have been classified in recent years as hazardous to health or the environment may require development of special handling facilities and procedures in order to permit their continued use. Nevertheless, it would be unrealistic to assume that all future replacements will exactly match the originals. There may arise compelling cost arguments for changes in specifications. In some cases it may be desirable to substitute materials with better aging characteristics. Whatever the justifications may be, the potential performance impact of any change proposals must be carefully weighed and experimentally verified wherever possible.

Performance Margins of the Primary

A primary yield that falls below the minimum level needed to drive the secondary to full output is the most likely source of serious nuclear-performance degradation. In many cases this minimum yield is known only from calculations of uncertain validity. The single production

---

[12] The importance and the challenges of maintaining the needed capabilities in the production complex have been stressed in both the FY2000 and the FY1999 reports to the Congress by the Panel to Assess the Reliability, Safety, and Security of the United States Nulear Stockpile (J. S. Foster [Chairman], et al., Washington DC, November 8, 1999 and February 1, 2001).

verification test was never able to explore all worst-case stockpile-to-target sequence conditions. This was also true of the development tests, which additionally seldom involved designs precisely identical to the production versions. Because primary yield margins in these weapons can be increased by changes that would not require nuclear testing, it is possible to use enhanced margins to provide insurance against minor aging effects and changes in material or process specifications arising in the refurbishment of the weapons. Margin enhancements, as recommended strongly in several JASON studies, would therefore constitute a significant additional confidence measure.[13]

Primary yield margins can be increased by appropriate changes specific to each stockpile system. These include changes in initial boost-gas composition, shorter boost-gas exchange intervals, or improved boost-gas storage and delivery systems. These modifications have been validated by nuclear test data for the appropriate systems, and they would not place burdens on the maintenance or deployment of the systems by the military. They would not require further nuclear testing, would help compensate for potential future uncertainties arising from remanufacturing, and are highly desirable where appropriate independent of a CTBT. We urge that these be done.

Non-nuclear Component Development and Manufacture

Since the non-nuclear components and subsystems external to the nuclear subsystem can be fully tested under a CTBT, it is possible to incorporate new technologies in these weapon parts as long as these can be shown not to have any adverse effect on proper functioning of the nuclear subsystem. In fact, it would be very costly to replace some non-nuclear components with exact copies as changeouts become necessary. If technologies involved in the non-nuclear components become prohibitively difficult to support with the passage of time because they are no longer utilized in the private sector, needed replacements can be based on current materials, technologies, and manufacturing processes.

This does require, however, the provision of adequate resources to provide not only the needed manufacturing capability and capacity but also for the associated engineering R&D and systems integration capabilities, on an ongoing basis. In many cases the component designs are complex, necessitating a lengthy development process and tight manufacturing controls. Safing devices, neutron generators, boost-gas subsystems, firing sets, radar fuzes, retarding parachutes, among many other active subsystems, must tolerate many years of storage and afterwards are expected to function flawlessly the first (and only) time that they are activated. A thorough understanding of possible failure modes during the entire stockpile-to-target sequence is essential. Because of the demanding specifications, a "science-based" approach to development and surveillance has long been a necessity.

Since a CTBT does not inhibit non-nuclear component development (with the possible exception of features related to radiation hardening) it introduces no new impediments to the recruitment of new scientific and engineering talent for such purposes. The principal requirement for program sustainability, especially at Sandia where most of this activity is centered, is a stable budget at a level adequate to maintain the diversity of facilities and scientific/engineering talent that are needed to perform the non-nuclear stewardship function in a competent manner. In addition, it will be essential to practice "change discipline"—the avoidance of changes that are not

---

[13] See, for example, JASON Report JSR-99-305, *Primary Performance Margins (U)* (McLean, VA: Mitre Corporation, December 1, 1999).

strictly necessary—because modifications, even outside the nuclear subsystem, can influence the entire weapon system.

Maintaining Nuclear Design Capabilities

Nuclear-weapon design activities are not prohibited under the CTBT, and preserving the capability to develop new designs—in case such are ever needed—is a stated goal of U.S. policy.[14] The use of ever more capable computational tools and more realistic material models to understand the relevant database from past nuclear tests, together with the use of advanced hydrodiagnostic techniques to study stockpile-related issues, is an important part of preserving this design capability, as is the engagement of new people in design activities while more senior designers are still available to share their insights. The associated design and evaluation expertise will aid in interpreting and perhaps anticipating foreign activities in nuclear-weapon development. We do not believe that nuclear testing is essential to maintaining these design and evaluation capabilities, even though such testing *would* be essential to certifying the performance of new designs at the level of confidence associated with currently stockpiled weapons.

Change-Control Discipline

It is conceivable that confidence in the performance of the nuclear subsystem might erode over time because of concerns over the cumulative effects of multiple small changes in materials and/or processes that may be introduced in the course of periodic refurbishment operations. The need for such changes is expected to arise very infrequently because of the inherently robust nature of nuclear components. The rate at which changes that individually are too small to have a significant performance impact could collectively cause unacceptable performance degradations is consequently very slow, perhaps measured in decades. Nevertheless, it is prudent to avoid the possibility of such problems, or at the very least push their onset beyond any time horizon of interest, by instituting a rigorous, highly disciplined, change-control process. Such a process must begin with a thorough documentation of the original design and manufacturing specifications. Any subsequent deviations from these specifications during stockpile maintenance and subsequent remanufacture of nuclear components must be minimized even if they are judged to be acceptable from a performance standpoint. Any deviations that are judged necessary must be thor-

---

[14] In 1994—before the 1996 signing of the CTBT—the Department of Defense conducted a Nuclear Posture Review (NPR). The relevant portion of the classified September 1994 Presidential Decision Directive, "Nuclear Posture Review Implementation," PDD/NSC 30 (S), is summarized in an unclassified paragraph on page 14 of the report of the Stockpile Assessment Team of the U.S. Strategic Command (Birely et. al., *Strategic Advisory Group Stewardship Conference Report (S)*, U.S. Strategic Command, April 2001) as follows: "The NPR contains infrastructure requirements for the Department of Energy to ensure high confidence in the enduring stockpile, namely: maintain nuclear weapons capability without underground testing or the production of fissile material; develop a stockpile surveillance engineering base; demonstrate the capability to refabricate and certify weapon types in the enduring stockpile; maintain the capability to design, fabricate and certify new warheads; maintain a science and technology base; ensure tritium availability; and accomplish these tasks with no new-design nuclear warhead production." This reiteration of the NPR requirements in a 2001 STRATCOM report indicates that they have not been superseded since the signing of the CTBT. We note, however, that the requirement to "maintain the capability to design, fabricate and certify new warheads" does not imply an intention to *exercise* that capability while a CTBT is in force. As indicated above, we believe it would not be possible to certify new-type nuclear subsystem designs for entry into the stockpile without nuclear testing—unless by accepting a substantial reduction in the confidence in weapon performance associated with certification up until now, or a return to earlier, simpler, single-stage design concepts, such as gun-type weapons. Thus, if not only design but also certification for entry into the stockpile of sophisticated new nuclear-weapon types were deemed necessary in the future in the national interest, and a CTBT were in force, we judge that the United States would need to withdraw from the treaty in order to conduct the nuclear tests needed for certification.

oughly analyzed and documented before adoption. The resulting audit trail should make it possible to include consideration of possible cumulative effects in judging the acceptability of any change proposal. In order to avoid the introduction of interference effects between nuclear and non-nuclear components, prudence dictates that a similar discipline be practiced in regard to any design changes or changes in location of non-nuclear components in proximity to the nuclear subsystem.

Priorities in Stockpile Stewardship

DOE's Stockpile Stewardship Program addresses all of the critical areas identified above, including plans for an aggressive "Stockpile Lifetime Extension Program" (SLEP). There remains the question of program balance. Since there will never be enough resources to pursue all relevant areas with maximum intensity, priorities become an important issue. We emphasize that the quality and quantity of people needed in the SSP constitutes the most important resource. This resource can become the limiting factor even if budgets are not a constraint.

Two somewhat related balancing issues, in particular, need careful consideration:

1) <u>Short-range versus long-range emphasis</u>: Each year since the moratorium the stockpile has been formally certified safe and reliable. Plans to maintain the stockpile in that condition for the foreseeable future can be carried out by the weapon complex as it now exists, augmented by appropriate near-term investments both in the laboratories as well as in the production plants. Simultaneously, efforts are under way to address problems that are anticipated to arise in the more distant future. Much of this work is aimed at achieving a much deeper and detailed scientific understanding of weapon phenomenology, especially as it relates to the physics of nuclear subsystem performance. It is imperative that a balance be maintained between the longer-range activities and the more immediate program needs. When judged in terms of their stewardship role, the goals and associated timetables for the longer-range efforts have sufficient flexibility that it should be possible to sustain these without sacrificing the short-term deliverables.

2) <u>Nuclear versus non-nuclear subsystem emphasis</u>: Understandably, the ongoing debate on stockpile safety and reliability has focused on possible consequences of a CTBT. What has not emerged from this debate is a clear recognition that safety and reliability are determined largely by the non-nuclear subsystems, and that these are not subject to CTBT constraints. Yet, it is the inability to test the nuclear subsystem under a CTBT that has stimulated increased Congressional support for the weapon program at a time when the program as a whole was suffering budget shortfalls. This creates the possibility that future budgets will become skewed in a direction that would make it difficult for the non-nuclear component and subsystem work to be performed at the laboratories and the production plants.

Concluding Remarks

Stockpile stewardship by means other than nuclear testing is not a new requirement imposed by the CTBT. It has always been the mainstay of the U.S. approach to maintaining confidence in stockpile safety and reliability. Based on the available evidence, we conclude that the measures outlined in the preceding sections are both necessary and sufficient for maintaining confidence in the continued safety and reliability of the enduring stockpile. These measures are independent of nuclear testing and are therefore unaffected by a CTBT.

An essential requirement is to safeguard the vitality of the DOE nuclear-weapon activities at the three laboratories, the production plants, and collaborating institutions. In the event that quantity replacements of major components of the nuclear subsystem should become necessary, prudence would indicate the desirability of extensive formal peer reviews. Evaluation of the acceptability of age-related changes relative to original specifications and the cumulative effect of individually small modifications of the nuclear subsystem should also be subject to periodic independent review. Such reviews, involving the three weapon laboratories and external reviewers, as appropriate, would evaluate potential adverse effects on system performance and the possible need for nuclear testing.

We judge that the United States has the technical capabilities to maintain confidence in the safety and reliability of its existing nuclear-weapon stockpile under the CTBT, provided adequate resources are made available to the Department of Energy's (DOE) nuclear-weapon complex and are properly focused on this task. Specific measures to bolster stockpile confidence include intensifying stockpile surveillance, strengthening manufacturing/remanufacturing capabilities, and increasing the performance margins of nuclear-weapon primaries. With no new weapons replacing older weapons, one can anticipate that the average age of weapons in the enduring stockpile will increase over time far beyond past experience. For this reason we regard these measures to be essential with or without nuclear testing or a CTBT. Refining computational understanding of the existing nuclear-test database and stockpile-related hydrodynamic experiments can play a key role in protecting the capability to produce new designs against the possibility that new weapon types were deemed needed in the future and a return to testing mandated to develop and certify them.

Some have asserted, in the CTBT debate, that confidence in the enduring stockpile will inevitably degrade over time in the absence of nuclear testing. Certainly, the aging of the stockpile combined with the lengthening interval since nuclear weapons were last exploded will create a growing challenge, over time, to the mechanisms for maintaining confidence in the stockpile. But we see no reason that the capabilities of those mechanisms—surveillance techniques, diagnostics, analytical and computational tools, science-based understanding, remanufacturing capabilities—cannot grow at least as fast as the challenge they must meet. (Indeed, we believe that the growth of these capabilities—except for remanufacturing of some nuclear components—has more than kept pace with the growth of the need for them since the United States stopped testing in 1992, with the result that confidence in the reliability of the stockpile is better justified technically today than it was then.) It seems to us that the argument to the contrary—that is, the argument that improvements in the capabilities that underpin confidence in the absence of nuclear testing will inevitably lose the race with the growing needs from an aging stockpile—underestimates the current capabilities for stockpile stewardship, underestimates the effects of current and likely future rates of progress in improving these capabilities, and overestimates the role that nuclear testing ever played (or would ever be likely to play) in ensuring stockpile reliability.

2

CTBT Monitoring Capability

The energy released by a nuclear test explosion generates signals that are potentially detectable by radionuclide sensors, by optical and other electromagnetic sensors (both ground-based and satellite-based), and by elastic wave sensors (infrasound, hydroacoustics, seismology).

From the point of view of a CTBT State Party considering clandestine evasion, all of these signals must be kept low enough to prevent detection and identification—or at least to inhibit attribution. Numerous signals are recorded on a daily basis from non-nuclear sources and numerous data streams need analysis from each sensing technology to explore the evidence of a possible nuclear test. The monitoring institutions need to develop confidence in the ability to detect, identify, and attribute.

An analysis of overall monitoring capability requires consideration of all available technologies, often in combination. What counts is the capability of the whole monitoring system. But inevitably, a detailed evaluation has to explore the separate techniques. Table 2-1 summarizes the way in which several different technologies contribute to monitoring four different environments (underground, underwater, atmosphere, and space). The first four technologies listed in Table 2-1 are used by the International Monitoring System (IMS) of the Comprehensive Test Ban Treaty Organization (CTBTO), headquartered in Vienna, Austria. The fifth technology (electromagnetic) uses sensors of many types, none of which are used by the IMS. They can include ground-based or space-based detectors of the characteristic flash of a nuclear explosion in the atmosphere or in space. Information on nuclear testing in any environment potentially can also be provided by signals intelligence. The sixth technology, satellite photography, now available commercially at 1-meter resolution (limited by U.S. government policy and not technology), can be used for remote examination of activity at and effects on sites on land and in the ocean.

Table 2-1 Contributions of key technologies to CTBT monitoring of different test environments

Monitoring Technologies	Underground	Underwater	Atmosphere	Near Space
Seismic	major	major	secondary	none
Radionuclide	major	major	major	none
Hydroacoustic	secondary	major	secondary	none
Infrasound	secondary	secondary	major	none
Electromagnetic	secondary	secondary	major	major
Satellite Imagery	major	major	secondary	secondary

Although there are synergies among these different CTBT monitoring technologies and all of them are needed, seismology is the most effective for monitoring against the underground testing environment, which is the one most suited to attempts at clandestine treaty evasion. From the viewpoint of a sophisticated evader, this environment is also the one most suited to evaluating the performance of nuclear-weapon components via explosive testing. We therefore give emphasis to seismic monitoring of the underground environment.

Starting in the late 1940s, the United States developed a capability to monitor atmospheric nuclear tests and was successful in detecting the first, unsuspected Soviet nuclear test in late August 1949 by routine air sampling over the Pacific Ocean. Over the next decade the system for air debris sampling and infrasound detection was developed and an initial network of seismic stations was established to monitor anticipated underground testing. The Limited Test Ban Treaty (LTBT) of 1963 (banning signatories from nuclear testing underwater, in the atmosphere, and in space) did not incorporate an independent international monitoring system, but depended on the Nuclear-Weapon States' independent national technical means (NTM), which were directed at keeping track of each other's nuclear programs and possible testing by Non-Nuclear-Weapon States. The international community at the time appeared satisfied that this system of NTM monitoring was adequate.

For the next decade the United States (and presumably the Soviet Union) improved its worldwide monitoring system with substantially improved seismic capabilities and a variety of satellite sensors to monitor atmospheric and space nuclear explosions. In 1974, the United States and the Soviet Union signed a bilateral Threshold Test Ban Treaty, banning underground nuclear tests of yield greater than 150 kt,[1] which involved extensive, close cooperation between the two states, and led to mutual understanding of the relationship between seismic magnitude[2] and yield of tamped underground nuclear explosions at their respective test sites[3]—a subject in longstanding dispute.[4] As an example of informal cooperation, in 1977 the Soviet Union called the attention of the United States to apparent South African preparations for an underground test in the Kalahari desert that had been missed by U.S. intelligence and that led to actions that prevented further activity at the site.

With international efforts to negotiate a comprehensive test ban in the mid-1990s, the international community was no longer satisfied to rely on the NTM capabilities of the Nuclear-Weapon States (primarily the United States) to monitor the treaty, but wanted it to be based on a truly internationally-operated system with information available to all parties. While the new system in many ways duplicates existing NTM capabilities, it adds significantly to those capabilities and makes the CTBT a genuinely international undertaking. This makes possible the establishment of a much denser network of monitoring stations and the international acceptance of challenge inspections. The treaty is designed so that both the international system and NTM can be used to carry out the monitoring function.

A significant difference between international monitoring efforts and monitoring by NTM is that the former must treat the whole world more or less equally, while the latter can concentrate on particular nations that are of concern to the United States. NTM can also focus on

---

[1] A kiloton is the energy unit usually used for specifying the energy released in a large explosion. Originally it was taken to be the energy released by a thousand tons of TNT, but a kiloton is now defined as a trillion calories (4.2×10^{12} joules).

[2] This is the Richter magnitude, based on the logarithm of the amplitude of seismic waves recorded at large distances. Different types of seismic waves are described in the following section. As a logarithmic scale, an increase in magnitude by one full unit implies an increase by a factor of ten in the amplitude of ground motion. An increase in seismic magnitude by 0.3 units corresponds to a factor of two in amplitude.

[3] A tamped explosion is one in which there is little or no space between the explosive device and the surrounding rock; and the device is detonated at sufficient depth so that all the gas and other by-products of the explosion are largely, if not completely, contained beneath the ground surface.

[4] This experience is useful in translating the capability to monitor the CTBT, expressed in terms of seismic magnitude, into the capability to monitor expressed in terms of the equivalent yield of a tamped underground nuclear test. Documentation from Russia, made available in recent years, provides added validation of this monitoring experience.

particular regions within the countries of concern. Monitoring nuclear testing by NTM has been a reality for more than 50 years, often with efforts that were concentrated on active nuclear test sites. It can be expected to be exceedingly difficult if not impossible for monitoring stations, operated on behalf of agencies with responsibility for U.S. NTM, to be installed in certain parts of the world of concern to the United States. But in many cases the international system can make installations in such regions. (It would be a setback to U.S. monitoring efforts if it were denied access to data from the International Monitoring System as a result of a U.S. failure to fulfill its financial obligations to the IMS or for other reasons.) We assume that investments in monitoring systems will continue to be made if evasion is regarded as feasible and of significant impact. In this connection, it is important that airborne radionuclide detection capabilities and specific satellite payloads be considered for their contribution to national security and not simply, as has sometimes happened, as a cost to an agency's budget.

In the sections that follow, we describe general aspects of all monitoring technologies, review current capabilities for monitoring the four different testing environments, describe what can be done with confidence-building measures and on-site inspections, and give our conclusions on monitoring capabilities overall.

General Aspects of All CTBT Monitoring Technologies

The CTBT will in practice be monitored by:
- the international CTBT Organization in Vienna, Austria;
- National Technical Means, which for the United States includes the Atomic Energy Detection System (AEDS) operated by the Air Force Technical Applications Center (AFTAC); and
- the loosely organized efforts of numerous institutions, acquiring and processing data originally recorded for purposes other than treaty monitoring, e.g. from regional and national networks of seismic and radionuclide sensors. Hundreds of institutions continuously operate thousands of seismometers; and seismically active regions of North America, Europe, Asia, North and South Africa, and the Middle East are now routinely monitored down to low magnitudes in order to evaluate earthquake hazards.

The international monitoring effort based in Vienna is specified partly by the CTBT text and partly by parallel agreements signed by countries specifying their commitments to the CTBT Organization. The CTBTO operates an International Data Centre (IDC) using data contributed by the radionuclide, seismic, infrasound, and hydroacoustic networks of the International Monitoring System. Figure 2-1 shows the location of more than 300 stations of the IMS, using different symbols for the 50-station primary seismic network, the 120-station auxiliary seismic network, and the 60-, 80-, and 11-station networks for infrasound, radionuclide, and hydroacoustic technologies, respectively. All these stations are to have a satellite link to the IDC in Vienna, some sending data continuously (e.g., the primary seismic network), others sending data only on request or at regular intervals.

Monitoring for underground nuclear explosions usually entails:
(a) detecting signals recorded by each sensor of a particular network;
(b) associating into a single group the various signals (from different sensors) that appear to be generated from a common source, often called an "event";
(c) estimating the location and time of that event and the uncertainty of the location estimate;
(d) identifying the nature of the event—whether suspicious or not in the context of CTBT monitoring; and

(e) attributing the event, if it is a nuclear explosion, to a particular nation.

The role of the IDC in event identification, is that of "Assisting individual States Parties ... with expert technical advice ... in order to help the State Party concerned to identify the source of specific events."[5] The responsibility of identification and attribution is left to each State Party. The IDC assists by carrying out a process called event screening, described below. The treaty mandates that each State Party maintain a National Authority to serve as the national focal point for liaison with the CTBTO and with other signatories. The U.S. National Authority for the CTBT would be a new entity.[6]

Much of the work of CTBT monitoring is routine, for example estimating the location of 50 to 100 earthquakes per day. In contrast, several years of experience with predecessors to the current IDC have shown about once or twice a year that some particular events have needed special attention to interpret their nature, including special efforts to acquire data from stations not operated by the IMS that recorded the event of interest. Examples include two mine collapses in 1995 (one in the Urals, the other in Wyoming); a small earthquake on August 16, 1997 under the Kara Sea near Russia's former nuclear test site on Novaya Zemlya; and two underwater explosions in August 2000 associated with loss of a Russian submarine in the Barents Sea.

Assessments of monitoring capability are typically made by evaluating routine procedures for detection and identification, since these are the methods that are first applied to an event of interest. It is necessary to make a prompt evaluation so that additional resources can be brought to bear on events deemed suspicious. The location and identification of events subjected to special study are more reliable, since they are based on more data and on methods of analysis that can be more complete and more sophisticated.

In this report we assess monitoring capabilities in two ways: first, against nuclear tests conducted in the manner typical of past practice by the Nuclear-Weapon States, with no effort to conceal signals; and second, against various evasion scenarios intended to reduce and/or mask the signals of a nuclear test, and thus to prevent detection, identification, and attribution.

There is a considerable degree of consensus among experts on the capability of various networks to monitor the first type of nuclear testing. This capability is significantly better than has commonly been believed. For monitoring against evasion scenarios, the issues are whether the proposed scenario is practical in view of the number of different ways that tests are monitored, and whether the scenario has utility for weapons development.

Concerning the step of attributing an identified nuclear test to a particular nation, procedures would differ somewhat depending on the environment in which the test was conducted. For the underground environment, there is the potential for long-lasting indications of the testing location (for example, a shaft or tunnel leading to a chamber with radioactive indicators of the explosion), whose coordinates may be estimated from seismic data followed up by identification of the site from satellite photos and other data, perhaps acquired as part of an on-site inspection. Attribution is likely to be more problematic for an underwater or atmospheric test, since a nation with a nuclear explosive could detonate it on a ship or a plane and the effects on the surrounding media would be more ephemeral. Though such a test would likely be detected and located, it might be attributed only with difficulty to the nation responsible. Measurement of the yield and other aspects of the explosion could be done to some extent through radiochemical analysis of one or more debris samples, and from the strength of seismic, hydroacoustic, and/or infrasound signals. Radionuclide collection may lead to identification of the type of nuclear explosive, in

---

[5] U.S. Department of State, "Protocol to the Comprehensive Nuclear Test Ban Treaty" 24 September 1996, pt. I, para. 20(c), http://www.state.gov/www/global/arms/treaties/ctb.html.

[6] Details of how this entity would operate have yet to be determined. The U.S. National Authority for the Chemical Weapons Convention is organized within the Department of State, and a similar arrangement is contemplated for the CTBT, with input from other agencies.

turn associated with the type of design used by a particular nation. To confidently evade attribution, a tester would need to believe that the United States, working with other nations, did not have the capability to track ships and planes in the vicinity of the test location, and would not intercept communications relating to the test. Attribution of a nuclear test in space is less of a problem, because of intensive efforts now made to track all objects launched into this environment.

Finally on this subject, we note that discussion of attribution is complicated when considering the possibility of a country that designs a nuclear explosive device and tests it on the territory of another country. The physical evidence of the test would point to the second country, which (if it provided assistance and was a State Party) would be a CTBT violation by the second country. Identification of the first country would most likely have to be made by NTM, including intelligence methods, or with the assistance of the second country.

Monitoring Underground Nuclear Explosions

Underground nuclear tests are primarily monitored with seismic and radionuclide signals. Important information can also be provided by satellite imagery.

Seismic signals are traditionally grouped into teleseismic waves and regional waves, depending on the distance at which they are observed. Teleseismic waves propagate either as "body waves" through the Earth's deep interior, emerging with periods typically in the range 0.3 to 5 seconds at distances greater than about 1,500 km, or as "surface waves" (analogous to the ripples on the surface of a pond) with periods of about 20 seconds. Because teleseismic waves do not greatly diminish with distance in the range from about 2,000 to 9,000 km, they are suited to monitoring a large country from stations deployed outside that country's borders. Teleseismic body waves are further subdivided into P-waves and S-waves. P-waves, which are the fastest-traveling seismic waves and are therefore the first to arrive, are particularly efficiently excited by explosions. Earthquakes tend to excite S-waves more efficiently. Teleseismic waves were the basis of most U.S. monitoring of foreign nuclear tests prior to 1987. Regional waves are of several types (including P-waves and S-waves), all propagating only at shallow depths (less than 100 km) with periods as short as 0.05 seconds, and they are regional in the sense that typically they do not propagate to teleseismic distances.[7] The word regional here carries the additional implication that such waves are dependent on local properties of the Earth's crust and uppermost mantle—which can vary strongly from one region to another. Regional wave amplitudes recorded up to about 1,200 km from a shallow source are typically larger than teleseismic wave amplitudes recorded at distances greater than 1,800 km, but regional waves are complex and harder to interpret than teleseismic waves. For sub-kiloton explosions, teleseismic signals can be too weak for detection at single stations and monitoring then requires regional signals.

The utility of regional waves is demonstrated by experience with monitoring the 340 underground nuclear tests now known to have been conducted at the Semipalatinsk test site of the Soviet Union in East Kazakhstan during the period 1961–1989. These explosions were mostly documented at the time by Western seismologists using teleseismic signals. But when archives of regional signals from Central Asia became openly available following the break-up of the Soviet Union, it became possible to detect and locate 26 additional nuclear explosions at this test site, most of them sub-kiloton, that had not been recognized or documented with teleseismic signals.

All modern seismographic stations operated to high standards (including all IMS stations)

---

[7] An exception is the Lg wave, consisting of shear wave energy trapped in the Earth's crust, and which under favorable circumstances can propagate to distances of several thousand kilometers.

use broadband sensors, with high dynamic range and digital recording of motions in three directions (up/down, North/South, East/West).[8] At many IMS stations, arrays of vertical component sensors are also operated. A seismographic array consists typically of between 5 and 30 sensors with short-period response, spaced over several square kilometers, and operated with a central recording system. Arrays provide the ability to detect smaller signals by taking advantage of the known correlation structure of signals and noises, and the ability to estimate the directions from which signals are arriving at the station by interpreting the time sequence at which signals reach individual sensors. The CTBT Protocol lists 50 primary station sites around the world (see Figure 2-1) at which either a three-component seismographic station or an array of seismometers is operated, sending continuous data in near-real time by satellite to the International Data Centre.[9] Much experience is now available from a network of similar size, operated since January 1, 1995, initially under the auspices of the Geneva-based Conference on Disarmament which negotiated the text of the CTBT, and later with endorsement from the CTBTO in Vienna. The associated data center, based near Washington, DC, became the Prototype IDC (PIDC) shortly after the CTBT was opened for signature in 1996, and served for 5 years as the test bed for many procedures that were taken up by the Vienna-based IDC in February 2000.

Data from the 50-station primary network provide the basis for event detection by the IDC and for initial estimates of source location. For events whose location and other attributes are likely to be better quantified by additional signals, the CTBT Protocol lists an additional 120 sites around the world at which either a three-component seismographic station or an array is to be operated. These stations, which constitute the IMS auxiliary network with locations also shown in Figure 2-1, record continuously; but their data are sent to the IDC only for time segments requested by a message from the IDC. Much experience with such an auxiliary network has been acquired by the Prototype IDC. During the 5-year time period from January 1995 to February 2000, the PIDC made available (at http://www.pidc.org) a number of reports of global seismicity derived from the primary and auxiliary networks. The most important report is the daily Reviewed Event Bulletin, typically published 3 to 5 days in arrears, listing the events that have occurred that day, including location estimates and their uncertainties, magnitudes, a list of the reporting stations for each event, and arrival times of detected seismic waves at these stations. About 50 events have been reported each day on average, with some wide variations, for a total of about 100,000 events in the first 5 years. The production of reviewed bulletins must be done combining all four monitoring technologies used by the IMS (seismology, hydroacoustics, infrasound, and radionuclide monitoring). In practice, normally all of the events reported each day by the IDC are either earthquakes or mining blasts. Nuclear explosions by France and China prior to their signing the CTBT in September 1996 were widely recorded and promptly characterized, as were nuclear explosions of India and Pakistan in May 1998.[10]

The transfer of IDC operations to Vienna in February 2000 has been associated with a significant change in practice, namely that the daily Reviewed Event Bulletin is no longer being made openly available to the research community. The IDC makes this bulletin (and all IMS data upon which it is based) available to National Data Centers, but restricts further dissemina-

---

[8] A "broadband" sensor can record teleseismic surface waves and body waves, as well as the much higher frequencies of regional waves.

[9] These data, as for all IMS data sent to the IDC, include information authenticating that the sensor and its data stream have not been compromised.

[10] In 1998, more than 60 IMS stations detected the Indian test of May 11 and the Pakistani test of May 28, while more than 50 detected the second, smaller Pakistani test of May 30. The Indian government announced that it also conducted two subkiloton tests on May 13 at 06:51 Greenwich Mean Time. No such tests were detected at the regional stations that had acquired large signals from the Indian test two days earlier. Because the May 11 explosion was detected at many stations, with signal/noise that in one case exceeded 1,000, it is clear that the existing seismic network was capable of detecting a very small Indian test if the test actually occurred. The seismic signals from any Indian test on the May 13 event were at least 500 times smaller than those of the May 11 event. There is no indication that evasion attempts played a role in concealing the later announced test. See B. Barker, et al., "Monitoring Nuclear Tests," *Science* 281 (1998): 1968-1969.

tion. The quality of IDC operations would be greatly enhanced in the long term by having a diverse community of data users. Many States Party, including the United States, have indicated their position that all IMS data should be made openly available without delay.

Seismic Detection Capability

Based on experience with noise measurements and station operations, the expected detection threshold of the CTBT primary seismographic network when completed is shown in Figure 2-2 in terms of contours of seismic magnitude. From events at this threshold and above, entailing detection of the first-arriving signal (always a P-wave) at three or more stations, a useful location estimate can be made. The auxiliary network does not contribute to the IMS detection capability shown in Figure 2-2, because as currently planned, data from this network are not available to the IDC unless requested for specific time segments—for example to enhance the data set collected from an already-detected event.

To interpret detection threshold maps such as shown in Figure 2-2 in terms of underground explosive yield we must use magnitude-yield relationships, which for tamped underground explosions typical of past experience are known to show some variability for different source regions, different depths for the explosion, and different propagation-path geologies. Table 2-2 lists yields as a function of decreasing magnitude for two representative magnitude-yield relationships in hard rock. The first (Y_1) is based on a relationship derived from extensive studies of Soviet underground nuclear testing at Semipalatinsk, Kazakhstan.[11] The second (Y_2) uses a relationship which is more appropriate for small explosions fired at a fixed depth.[12] Table 2-2 also indicates approximate yield values that are representative of hard rock explosions with magnitude 3.5 and smaller. Figure 2-3 uses these approximate yield values to indicate yield detection thresholds of the IMS primary seismic network for major parts of Europe, Asia, and Africa. Thus Figure 2-3 is an expanded view of part of the global map in Figure 2-2, but with magnitude contours now interpreted as approximate yields (via Table 2-2) for small tamped explosions in hard rock.

Table 2-2 Yields (in kilotons) of small nuclear explosions tamped in hard rock that correspond under different assumptions to a set of decreasing seismic magnitudes. (The approximate values listed in the final column are used in Figure 2-3 to indicate detection thresholds in terms of nuclear yield for tamped explosions in hard rock.)

Seismic Magnitude	Y1	Y2	Approximate Y
3.50	0.055	0.125	~0.10
3.25	0.025	0.071	~0.06
3.00	0.012	0.040	~0.03
2.75	0.005	0.022	~0.02
2.50	0.003	0.013	~0.01

Yields at a given seismic magnitude can be different from those listed in Table 2-2 to the extent that an explosion is not tamped and/or not in hard rock. Thus, explosions in clay or water

---

[11] Magnitude = 4.45 + 0.75 log Y (in kt): see for example F.Ringdal, et al., "Seismic Yield Determination of Soviet Underground Nuclear Explosions at the Shagan River Test Site," *Geophysical Journal International* 109 (1992): 65-77. The slope here, with value 0.75, is less than unity because larger shots are usually fired at greater depth, and hence couple less efficiently into seismic energy. This manitude yield relation applies to the southern Novaya Zemlya test location and (with slightly lower intercept of about 4.25) to the northern Novaya Zemlya test location.

[12] Magnitude = 4.4 + log Y (in kt).

couple more efficiently into seismic signals so that the yield producing a given magnitude can be 10 or more times smaller than listed. Explosions in soft rock couple less efficiently and the yield producing a given magnitude can be 10 or more times larger than listed. An explosion in an underground cavity also couples less efficiently, raising the question of evasive testing, discussed below.

Figure 2-3 is important as indicating that the detection capability of the IMS primary seismic network is very significantly better than 1 kt for nuclear explosions conducted in a fashion typical of past underground nuclear testing (i.e., tamped and without efforts to reduce signals). For most of Europe, Asia, and Northern Africa, the detection threshold is down in the range from 30 to 60 tons in hard rock. The IMS was intended (but not formally specified) to support monitoring down to about one kiloton. During most of the more than 20 years of planning and building up the seismic components of this network, monitoring experience was almost all based on teleseismic waves although it was expected that regional waves would be superior for monitoring at lower yields. The IMS was designed with station spacing sufficiently close to enable use of regional waves, but it is only now becoming apparent from results such as those of Figures 2-2 and 2-3 that regional waves enable monitoring to be done well below 1 kt.

Association of Signals

A consequence of the low thresholds indicated in Figures 2-2 and 2-3 is the large number of seismic events that are detected and therefore require some analysis. Each year, somewhat more than 7,000 earthquakes occur worldwide with magnitude greater than or equal to 4, and about 60,000 with magnitude greater than or equal to 3. While chemical explosions having magnitude greater than or equal to 4 are rare (a few per year, if any), there are probably on the order of a few hundred per year worldwide with magnitude greater than or equal to 3, and many thousands per year at smaller magnitudes that are detectable with stations close enough.

The problem then arises of sorting through the tens of thousands of signals each day that will be detected from network analysis of array data and three-component station data, and collecting together all the signals that are associated with the same seismic source.[13] In the work of assembling sets of detections common to the same event, array stations in principle have the advantage, over three-component stations, of permitting determination of the direction of the source from a station at which there is a detection.

Associating signals correctly to the underlying seismic events is currently the most challenging software problem at the IDC. This challenge has been met in recent years for detections from the primary network. It could also be met using a larger network (for instance, with the addition of the auxiliary network, if it were enabled to report continuously), which would drive detection thresholds down to lower magnitude.

---

[13] Each seismic source generates several different waves, each arriving at a different time. Signals from different events can therefore overlap in time at each station. Effective methods of sorting through the interspersed detections have been found, but in practice 40 percent or more of the detections recorded on a given day have remained unassociated at the PIDC. The unassociated signals are typically from events of such small size that they are not detected by three or more stations, and have magnitudes lower than those contoured in Figure 2-2. These unassociated detections from low magnitude events are an unavoidable consequence of operating a very sensitive network.

Location of Seismic Sources

In practice, location estimates are needed for all the events for which a set of detections can be associated, since an interpretation of the location (including the event depth) is commonly used to reach preliminary conclusions on whether an event could possibly be an explosion. It can be crucial to know if an event is beneath the ocean, or possibly on land; if it is more than 10 km deep, or possibly near the surface; if it is in one country, or another. The source-station distance and azimuth can be roughly estimated from data at a single station, such estimates being much better in the case of an array station than for a three-component station. The accuracy of a location estimate is best characterized with a confidence interval, and typically detections at three or more stations are needed for confidence intervals to be calculated.

The CTBT Protocol, Part II, paragraph 3, states that "The area of an on-site inspection shall be continuous and its size shall not exceed 1,000 square kilometers. There shall be no linear dimension greater than 50 kilometers in any direction."

This condition presents a challenge for those who may have to estimate the location of an event that could become the basis of an on-site inspection request. The challenge is especially difficult for small events, when the estimate may have to be based upon regional seismic waves alone. Regional waves can be used to make accurate location estimates, but only in regions that have previously been well studied.

The treaty language limiting the area of an on-site inspection has been taken by the monitoring community as setting the goal of locating seismic events with 90 percent confidence intervals no greater than 1,000 square km in area. This goal is being actively pursued. Methods first developed and applied to seismic events in Northwestern Europe and North America entail calibration of IMS stations in these regions, by developing station-specific information on the travel-times of regional waves arriving from sources at different distances and azimuths. These methods are now being extended to IMS stations in North Africa, the Middle East, and throughout Europe and Asia.

Identification of Seismic Sources

Once the detections from a seismic event have been associated and an accurate location estimate has been obtained, the next step in monitoring is that of event identification. As noted above, the IDC's role in event identification is limited to providing assistance to treaty signatories. The CTBT Protocol indicates that the IDC may apply "standard event screening criteria" based on "standard event characterization parameters." To screen out an event means that the event appears not to have features associated with a nuclear explosion. If the IDC can screen out most events, then treaty states can focus attention on the remainder. For example, an event may have its depth estimated with high confidence as 50 km. Such an event would be screened out by a criterion that eliminates events confidently estimated as deeper than 10 km.

The most successful discriminants are based on the location (including the depth), and analysis of the amplitude ratio between different types of seismic wave. Location is important as a discriminant, because earthquakes are very rare (or totally unknown) in most regions of the world—for example for most of the territory of the Russian Federation. Any seismic signal from such an aseismic region will therefore attract attention and require careful scrutiny. At the opposite extreme, represented for example by much of China, earthquakes are common and there is a large population of previous earthquakes against which to compare the signals of a new event. An archive of such earthquakes will gradually be built up by IMS stations as that network becomes established. Archives already exist for numerous non-IMS stations.

Based on practical experience with teleseismic signals, there is an extensive literature on methods of discrimination between earthquakes and explosions. For example, comparison of teleseismic body-wave and surface-wave amplitudes has empirically been shown to provide an excellent discriminant. Surface waves from a shallow earthquake with the same body-wave strength as an explosion are typically 6 to 8 times larger than surface waves from the explosion. But this method usually cannot be used for events much below magnitude 4 because teleseismic surface waves are then typically too small to detect.

For teleseismic signals, a rough rule is that identification thresholds are about half a magnitude unit above detection thresholds, where detection thresholds are interpreted as in Figure 2-2 (i.e., a requirement for detection at three or more stations in order to provide a useful estimate of the location).

Practical experience with discrimination using regional signals is more limited, but growing, and the best methods are based on the use of spectral ratios of two different regional waves. In practice, the particular spectral ratio that is most effective for discrimination can be different for different regions. Discrimination has been demonstrated down to magnitude 3 for many regions. This capability requires access to data of high quality for the event of interest, and adequate data sets of previous explosions and earthquakes against which that event can be compared. Studies now underway are building up confidence in discrimination based on regional waves, for different areas of interest. In practice, a particular "problem event" in or near a region of interest has sometimes spurred the necessary special studies to enable discrimination to be done to low magnitudes using regional waves. Such was the case for a small seismic event (magnitude about 2.5) that occurred near the Novaya Zemlya test site on December 31, 1992, which after much effort lasting several months led to an improved understanding of regional waves from sources near this site, and identification of the event as a very small earthquake.

For regional waves, it appears that identification thresholds may not be significantly different from detection thresholds. This conclusion is based on the fact that for detection thresholds defined as in Figure 2-2, with detection required by at least enough stations to provide a useful location estimate, it is likely that one or more of the stations will have signal levels high enough to enable the measurement of spectral ratios suitable for discrimination. In the context of the IMS network, with detection thresholds set solely by the primary stations, auxiliary stations provide additional sources of data for measuring spectral ratios. And in the case of an event of sufficient interest to require going beyond routine analysis based only on IMS data, supplementary data from stations operated for purposes other than treaty monitoring can be used.[14] Vast regions of Europe, Asia, North America, North and South Africa, and the Middle East are currently being monitored for earthquakes down to magnitude 3 or lower by networks operated for purposes unrelated to the CTBT. There are literally thousands of seismometers deployed around the world. Their numbers are growing, as is the trend of data quality and of products derived from the data. Basic reasons for these improvements are the quality of modern hardware and software, which are replacing earlier systems; a substantial and growing infrastructure of geophysical research into Earth structure and the science of earthquakes; and general concern over earthquake hazards.

For small regions that have been intensively studied, such as former nuclear test sites, the identification threshold can be even lower than the three-station detection threshold, since just one station at regional distances can provide discrimination capability if the signals at this station are of high enough quality, even though the usual standards for obtaining an accurate location are not met. Of course it is not good to rely upon a single station (especially if it is on the territory

---

[14] Such stations cannot always be relied upon, their data may not be obtainable for several days, and if stations are available they must be evaluated and perhaps calibrated in an *ad hoc* fashion. But practical experience with several problem events has demonstrated that useful data from supplementary stations can often be found. In some cases they have provided the best available data for such events.

of a country where suspicious events occur), and a single station may not be regarded as adequate for building the case that a particular event should be investigated with an on-site inspection. But the overall view of monitoring capability should be developed in the context of the whole system, which may include satellite data and other non-seismic methods of monitoring, once a single (seismic) station provides evidence of the possible occurrence of a small explosion.

A preliminary result from screening studies at the PIDC is that more than 60 percent of events with magnitude greater than or equal to 3.5 are routinely screened out on the basis of a conservative interpretation of their regional-wave spectral ratios. In practice, objective screening based upon such a discriminant may be developed over a period of time and routinely applied at the IDC in ways that may have to be fine-tuned slightly differently for different regions.

Some successful methods of event identification use a combination of seismic and other data. For example, about 70 percent of all earthquakes occur under the ocean and many of these can easily be recognized as earthquakes on the basis of an absence of strong hydroacoustic signals.[15]

Mine blasting can result in seismic events smaller than magnitude 4 that may be detected and located by the IDC. These events can appear similar to small nuclear explosions using standard criteria such as depth estimates and weak surface waves. For a chemical explosion fired as a single charge at a depth sufficiently great to contain all the fractured material, there is no difference seismically from a small nuclear explosion, and no indication (as to the type of explosion) from ground deformation. But mine blasts above a few tons are in almost all cases executed for the commercial purpose of fracturing large amounts of rock, and as such they routinely consist of numerous small charges which are fired in a sequence of delays. This practice is sometimes called "ripple firing." The resulting seismic source is therefore spread out in space or time, or both, and can be identified as different from a single-fired source by use of high-frequency seismic signals. Another potential discriminant of mine blasting is the infrasound signal which can be generated by explosions used in surface mining, and which would not be expected from a small underground nuclear explosion unless it vented significantly—which, in turn, would give a very strong radionuclide signal.

Radionuclide Releases From Underground Explosions

Experience indicates that underground testing has often given rise to radioactive releases into the atmosphere, and can therefore potentially be detected by the IMS radionuclide monitoring network, which is primarily directed at monitoring atmospheric nuclear explosions (see below). Recent Russian papers documenting Soviet nuclear testing state that all underground tests at Novaya Zemlya and about half the underground tests at the Semipalatinsk test site in Kazakhstan resulted in release of radioactivity.[16] Radionuclide detection plays a significant role in the remote detection and identification of nuclear tests. A decoupled blast in the 1 to 5 kiloton region, for example, could vent significant quantities of debris proportional to the nuclear yield, irrespective of the decoupling of the seismic signal. On-site inspection, discussed below, provides the opportunity to detect, near to the test site, radioactivity that may have been generated by a nuclear explosion in amounts too small to be detectable at greater distances.

---

[15] If a seismic event is located with confidence in an ocean area, it can only be an explosion if the device was set off in a hole drilled into the ocean floor, or if it was set off in the water itself. The logistics of sub-oceanic drilling to sufficient depths for containing a nuclear explosion are so formidable as to rule out this possibility in most areas. The simplest way to screen an oceanic seismic source is then to examine hydroacoustic data, noting that an explosion in the water would generate very large sound waves within the water. An absence of hydroacoustic signals can therefore screen out, as earthquakes, all seismic sources confidently located in ocean areas.

[16] For example, V.N. Mikhailov, et al., *Northern Test Site: Chronology and Phenomenology of Nuclear Tests at the Novaya Zemlya Test Site* (July 1992).

Monitoring Against Underground Evasion Scenarios

The concept of evasive testing has been extensively explored since 1959, when the best known evasive method (cavity decoupling) was first proposed, and presented by the United States at trilateral CTBT negotiations with the United Kingdom and the Soviet Union. Some of the hypothesized methods of evasive testing are intended to reduce specific types of signals from a nuclear explosion; some are intended to mask or disguise the signals by combining them with the signals from a non-nuclear source such as chemical explosions used in mining; and, as noted earlier, some methods focus on evading attribution.

The energy released by a nuclear explosion, at the level of a kiloton or more, is much greater than can be physically contained in practice by man-made structures above ground. Therefore, evasive testing in the underground environment is generally regarded as the most serious challenge to monitoring efforts.

In the era of monitoring underground nuclear tests principally by use of teleseismic signals, serious consideration had to be given to the concept of hiding a nuclear explosion by testing shortly after a large earthquake. This possibility, which faces the difficulties of preparing a test and waiting months or longer for a suitable earthquake, and of identifying and evaluating such an earthquake and carrying out the test within minutes of the earthquake's occurrence, is rendered far harder to execute by the fact that today's monitoring systems record regional seismic waves. These waves in practice are not reliably obliterated by teleseismic signals, even from the largest earthquake.

The most serious methods of evading detection and identification of an underground nuclear test are decoupling in large cavities and masking by mine blasting.

Decoupling

If a nuclear explosive device is tested in a large enough cavity, constructed deep underground, then almost all the explosive energy goes into pumping up the gas pressure within the original cavity. This contrasts with a tamped explosion, for which the energy goes into non-elastic processes such as melting and crushing rock, thus creating a new cavity and strong seismic signals. An explosion in a previously constructed cavity therefore decouples much of the energy that would have gone into seismic signals, reducing them by a so-called decoupling factor which can be as high as 70. Such an explosion is said to be "fully decoupled" if the original cavity walls are not stressed beyond their elastic limit. In practice with chemical explosions, decoupling factors of 10 to 30 have been typical.

The United States carried out a decoupled nuclear test of 0.38 kt in 1966, in the cavity created years earlier in a salt dome by a much larger nuclear explosion, and demonstrated that a decoupling factor of about 70 for a small yield is indeed possible. At a fixed depth the volume of the cavity has to increase in proportion to the yield to achieve the same decoupling factor, but as yield increases so do the practical difficulties of clandestine cavity construction and radionuclide containment.[17] The Soviet Union carried out a partially decoupled test of about 8 to 10 kt in 1976, in a cavity (in salt) of mean radius 37 m (sufficient to fully decouple about 3 kt). But this event was decoupled only by a factor of about 15, and it had magnitude about 4.1. The resulting seismic signals were picked up teleseismically as far away as Canada, and according to news reports at the time were promptly identified as originating from a decoupled nuclear test.

---

[17] In salt at a depth of around 1 km, the spherical cavity required for full decoupling of a 1-kt explosion has a radius of about 25 m and a surface area of about 8,000 square meters. This area, which is increased for a non-spherical cavity, would be exposed to a gas pressure of around 150 atmospheres. At a fixed yield, the volume of the cavity must be increased as the depth is decreased, in order to achieve containment.

Aspherical cavities, which are easier to construct than spherical cavities of the same volume, can also achieve high decoupling factors, but they also weaken the cavity walls by concentrating stress, making it more likely that radionuclides will be released.[18]

There is no practical experience with cavity decoupling of a nuclear explosion in hard rock, which is much more common than suitably large salt deposits. Hard rock in practice contains cracks on many length scales, and there is no experience with containing radionuclides in material that is intrinsically cracked and exposed to high gas pressures over large areas (tens of thousands of square meters). Containment is itself a highly technical subject, built up from empirical experience and learning from mistakes (for example at the Nevada test site). But key containment techniques relied upon for tamped explosions are not available for a decoupled explosion.[19]

Evaluation of the cavity decoupling scenario as the basis for a militarily significant nuclear test program therefore raises a number of different technical issues for a country considering an evasive test:

- Is there access to a region with appropriate geology for cavity construction?
- Is there such a region which is suitably remote and controllable, and that can handle the logistics of secret nuclear weapons testing?
- Can cavities of suitable size and shape and depth and strength be constructed clandestinely in the chosen region, hiding the material that has to be removed from below ground?
- Can the limited practical experience with nuclear tests in cavities in salt, and very low-yield chemical explosions in hard rock, be extrapolated to predict the signals associated with nuclear testing in cavities in hard rock?
- Can a decoupling factor as high as 70 be attained in practice for yields significantly higher than subkiloton?
- Can those carrying out the decoupled test be sure that the yield will not be larger than planned, and thus only partially decoupled?
- Can the site be chosen to avoid seismic detection and identification, given the detection thresholds of modern monitoring networks and their capability to record high-frequency regional signals?
- Can radionuclides be fully contained from a decoupled explosion?
- Can nuclear explosions of large enough yield be carried out secretly, and repeated as necessary, to support the development of a deployable weapon?
- Can secrecy be successfully imposed on all the people involved in the cross-cutting technologies of a clandestine test program, and on all the people who need to know of its technical results?

These questions represent layers of difficulty with the cavity decoupling evasion scenario, going outside practical experience in a number of areas. Most questions have been the

---

[18] Stable non-spherical cavities are easier and cheaper to construct than stable spherical cavities of the same volume, and it has been demonstrated for small chemical explosions that non-spherical cavities can provide approximately the same decoupling factor as that for the spherical cavity. But for a non-spherical cavity there is a significantly greater risk of radionuclide releases because (a) the surface area is greater (than for a sphere of the same volume), and so a greater chance of the pressurized gas encountering a crack in the walls, and (b) part of the walls must be significantly nearer the device than for a device at the center of an equivalent sphere. It is this fact that places a limitation on the aspect ratio of a nonspherical cavity, because of the possibilities for ablation shock (in the case of nuclear explosions, but not chemical explosions). Such a method for transferring energy into the walls can lead to wall damage promoting radionuclide release. In addition it can lead to non-elastic deformation and hence larger seismic signal generation and a smaller decoupling factor (for a nuclear explosion).

[19] See, for example, J. Carothers, et al., *Caging the Dragon: The Containment of Underground Nuclear Explosions* (DNA Technical Report 95-74, 1995).

subject of extensive technical analyses. For some questions, the answer is clearly affirmative.[20] For other questions, there are experts in particular fields who confidently assert that particular problems can be overcome.[21] But the real issue is whether the scenario as a whole can be made to work.

Although it would appear that an uppermost bound of 7 kt (evasive) is implied by detection capability at the 100-ton level or better (tamped; see Figure 2-3), we find this 7-kt value is not plausible because of the numerous obstacles that must simultaneously be overcome. It is the whole monitoring system that has to be addressed by a potential evader. Signals would have to be hidden from the IMS, from NTM, and from numerous sensors operated for purposes unrelated to treaty monitoring. In many instances, a series of nuclear tests would be necessary to achieve a tester's objectives. For reasons such as these, it has been difficult to reach a consensus in the monitoring community on a particular value of yield that all can agree is the maximum available to a potential evader—a yield above which a test would be reliably detected and identified by some particular component of the monitoring system. Rather, the basic issue is whether the cavity decoupling scenario itself is plausible given the difficulties and risks it entails. Only nations already possessing extensive knowledge of nuclear explosive devices and with access to suitable geologic sites could expect to address the difficulties of this type of testing.

It has been claimed that cavity decoupling is plausible for evasive testing not just at the level of 1 or 2 kt, but at 20 kt, and even as high as 50 kt. Ignoring completely the challenges of cavity construction and radionuclide containment, such explosions (even if fully decoupled by a factor of 70) would have seismic signals far above the detection levels contoured in Figure 2-3.

Accepting the possibility of a cavity decoupled test, we conclude that such an underground nuclear explosion cannot be reliably hidden if its yield is larger than 1 or 2 kilotons.

Mine Masking

Another proposed evasion scenario is the use of mining operations and large chemical explosions to mask or disguise an underground nuclear explosion. A country considering this approach would need to face some of the general problems outlined above for a cavity-decoupled nuclear test, including containment of radionuclides, imposition of secrecy on many people, and the capability of seismic monitoring.

As for mine masking, chemical explosions in mines are typically ripple-fired and thus relatively inefficient at generating seismic signals compared to single explosions of the same total yield. For a nuclear explosion that is not also cavity-decoupled to be hidden by a mine explosion of this type, the nuclear yield could not exceed about 10 percent of the aggregate yield of the chemical explosion. A very high-yield, single-fired chemical explosion could mask a nuclear explosion with yield more comparable to the chemical one, but the very rarity of chemical explosions of this nature would draw suspicion to the event. Masking a nuclear yield even as large as a kiloton in a mine would require combining the cavity-decoupling and mine-masking scenarios, adding to the difficulties of cavity decoupling already mentioned.

A confidence-building measure specified in the CTBT Protocol is an attempt to address this issue: "… each State Party shall, on a voluntary basis, provide the Technical Secretariat with notification of any chemical explosion using 300 tonnes or greater of TNT-equivalent blasting material detonated as a single explosion anywhere on its territory, or at any place under its juris-

---

[20] Thick salt deposits are known to exist for example in certain regions of Russia and the United States. Techniques have been developed for hiding material removed from underground.

[21] Large cavities in salt can be constructed by pumping in water and pumping out brine; and containment can be improved by going to greater depths—though cavity construction is then more difficult and subsequent work in the cavity is more dangerous.

diction or control." The Protocol adds that: "Each State Party shall, on a voluntary basis, as soon as possible after the entry into force of this Treaty provide to the Technical Secretariat, and at annual intervals thereafter update, information related to its national use of all other chemical explosions greater than 300 tonnes TNT-equivalent." Execution of this voluntary measure will amount to a significant effort, in that the mining industry uses megatons of chemical explosive each year (2 megatons/year in the United States alone).

If there were a nuclear explosion in a mine, and a challenge was issued to which the reply was "those signals came from a single-fired chemical explosion," pursuing the matter could be problematic insofar as, under the CTBT, information on chemical explosions is voluntary and cannot easily be checked. The basis for obtaining support for a formal on-site inspection could be difficult to develop. Again, however, the issue has to be addressed in terms of the size of a nuclear explosion that could plausibly be hidden in this way. Its magnitude would have to be somewhat lower than the biggest chemical explosions, and thus capped at about magnitude 3.5. Such yields if tamped are quite small (see Table 2-2). And so this scenario as a whole must be combined with other evasive measures, in order to reach the kiloton level.

There are two types of technical effort that can be carried out to help monitor mining regions for compliance with a CTBT. The first of these is installation of nearby seismographic stations that record digitally at high sample rates. Such data can distinguish between single-fired explosions and the multiple-fired explosions typical of mining operations. The second is provision of technical advice to mine operators so that they execute their blasting activities using modern methods of delay-firing—which have economic advantages, as well as enabling the blasting of rock in ways that do not make the large ground vibrations (and strong seismic signals) typical of old-fashioned methods of blasting.[22]

For those specific locations where mine masking remains a concern, this could be addressed with installation of infrasound and radionuclide monitoring equipment, and with site visits. Though such monitoring would be voluntary under the CTBT, it could be enhanced via bilateral agreements and/or by multilateral agreements between neighboring countries in a particular region. Such voluntary compliance would help in proving that the observed event was not a nuclear explosion.

Methods For Improving Seismic Monitoring Capability

The detection capability described in Figures 2-2 and 2-3 underlies much of the discussion above, and it is around 100 tons. If it is deemed necessary to achieve even better levels of monitoring treaty compliance (to lower yields and with greater confidence at the 100-ton level), there are a number of ways to make improvements.

(1) Capability could most simply and cost-effectively be improved by including continuous reporting from the 120 stations specified as the IMS auxiliary network. The detection capability of the IMS at present is based solely on the primary network, because continuous examination of seismic data is done only for these stations. Auxiliary stations are already recording continuously, and they already have satellite links to the IDC. If their data along with data from the IMS primary seismic network were continuously examined for detections, the monitoring thresholds shown in Figure 2-2 (three station detection capability) would drop generally by about 0.25 magnitude units in Europe, Asia, and North Africa, and by about 0.5 magnitude units in

---

[22] Both of these approaches are developed in National Research Council, *Seismic Signals from Mining Operations and the Comprehensive Test Ban Treaty: Comments on a Draft Report by a Department of Energy Working Group*, (Washington, DC: National Academy Press, 1998). The NRC recommended an emphasis on the first approach. The DOE working group had recommended the second approach.

some regions (such as Iran, for which the detection threshold would drop from about 3.25 to about 2.75).

(2) More generally, detection capability can be improved by any augmentation of the IMS primary seismic network, to the extent that the additional data streams are continuously examined for detections, along with IMS data streams. Augmentation can be done to improve the monitoring of areas of particular interest, for example regions of thick salt, and mining regions, to the extent there is concern over evasive testing in these areas and therefore a need to monitor them to lower magnitudes.[23]

(3) In cases where a small event is suspected at a location almost the same as that in which a larger event has been documented, then correlation analysis can sometimes provide good detection of the smaller event, even though its signals are not directly apparent. Though now a specialized and little-used technique, it has proven useful for detecting aftershocks (for example of the Kara Sea earthquake of August 1997, near the Novaya Zemlya test site), and can be used to search for very small explosions near the site of recorded larger explosions. In future decades, this may become the method of choice for detection of very small events in areas for which an archive of previously recorded events becomes available.

(4) Instead of searching for detections in the continuous data stream from each IMS station separately, the data streams can be added together from different stations with an appropriate time delay, and then subjected to continuous search.[24] This method, known as Threshold Monitoring, enables continuous measurement of the maximum magnitude at which a seismic event could have occurred at a particular location and a particular time. To the extent that there is no seismic activity in the broad region, this maximum magnitude can be very low. To the extent there is activity anywhere in the region, it can raise the maximum magnitude at the site of interest even if the activity did not occur at that location.[25]

The Threshold Monitoring method is well-suited to monitoring aseismic regions down to significantly lower magnitude than conventional single-station detection thresholds. In application to Novaya Zemlya, the NORSAR organization in Norway (which pioneered the Threshold Monitoring method) has demonstrated that the maximum magnitude at which any seismic activity could have occurred is lower than 2.5 for most of the time. But this maximum magnitude can occasionally rise above 2.5 on some days, due to seismic activity in the broad region of the Barents Sea and other areas. On other days, this magnitude is closer to 2, all day. On such days, the indication from Table 2-2 relating magnitude and yield (with all the assumptions this entails) is that Novaya Zemlya can be monitored down to below 10 tons (tamped). The Threshold Monitoring method can be applied in near-real time if data are promptly transmitted to an analysis center, or with a delay to the extent that data become available from a relevant station at a later date.

(5) Numerous seismic stations operated for purposes unrelated to treaty monitoring are not currently used in ways that enable detection thresholds to be lowered. Typically, either the data are not routinely examined at each station for small amplitude detections, or such detections are not routinely reported to any international data centre. The International Seismological Cen-

---

[23] Augmenting the already sensitive IMS primary seismic network as described in (1) or (2) would decrease the number of unassociated signals reported by the primary network alone (see footnote 13). The additional events detected at three or more stations would be of low magnitude, presenting challenges for event identification; and yet more events at even lower magnitude would still generate unassociated signals in large numbers with the augmented network. In general, augmenting a network improves monitoring capability by driving the monitoring threshold downward, but it also increases the number of unassociated signals below the threshold.

[24] The appropriate delays are different (for a given set of stations) for different source locations and for different seismic waves. Numerous different sets of delays can be selected, one for each source location, to seek evidence of small events at each source location. Knowing the noise levels at each station, it is then possible to state, for each source location, that if a seismic event occurred at a particular time, it must be below a certain magnitude.

[25] To find out where specific activity occurred, it is necessary to find detections at single stations, preferably three of them or more as in Figure 2-2.

tre (which at present receives detection information from about 3,000 stations) is potentially capable of serving the necessary role.[26] But most station operators around the world ignore the ISC; and those that do contribute detections often do not contribute them for small events. If, however, an organizing principle were agreed to by non-IMS station operators, to develop a joint bulletin describing all seismic activity down to about magnitude 3 on continents, there would be significant benefits for treaty monitoring.

Monitoring Underwater Nuclear Explosions

Underwater nuclear tests can be monitored by hydroacoustic, seismic, and radionuclide signals.

Hydroacoustic signals in oceans travel to great distances within a waveguide, sometimes called a "sound channel," so that explosions of only a few kilograms yield in most regions of the deep oceans are readily detected at distances of thousands of kilometers.[27] Such sources can be identified as explosive by detecting the presence of a characteristic bubble pulse in the recorded signal, caused by gases at the source expanding and contracting. Because hydroacoustic waves also couple efficiently into seismic waves at the ocean bottom (and vice versa), the underwater medium can be effectively monitored by a combination of hydroacoustic and seismic networks. Monitoring sound waves in the oceans is a well-advanced discipline, primarily as a result of investments in acoustic systems to detect submarines.

The hydroacoustic component of the IMS includes six hydrophone stations, each deployed in the ocean with signals sent by sea-floor cable to a nearby island for subsequent transmission by satellite to the IDC. It also includes five so-called T-phase seismic stations deployed on islands in five countries.[28] By itself this component of the IMS is not capable of locating a nuclear explosion with the desired precision. However, hydroacoustic data are useful for discrimination purposes (see footnote 15, on seismic discrimination of oceanic earthquakes) and will support seismic data used for location purposes.

Figure 2-4 shows that the yield threshold is down to just a few kilograms for most oceans in the Southern Hemisphere. Almost all the world's ocean basins will be monitored down to better than one ton.

Although coupling from explosive energy to oceanic waveguide is very efficient for the deep oceans, in the case of shallow oceans it is possible for horizontally-traveling acoustic energy to be blocked from reaching the waveguide, and hence from reaching distant hydrophones. But for explosive sources in shallow water, downward-traveling acoustic energy converts well to seismic waves at the ocean floor that can then propagate in the solid Earth. This coupling was demonstrated by two explosions in the Barents Sea, which is located north of Norway, on August 12, 2000. These explosions resulted in puncturing the hull of the Russian submarine *Kursk*. The second and larger explosion, in water about 100 m deep, was well recorded by several IMS seismic stations (both regionally and teleseismically at stations in Russia, Alaska, and Canada), and by non-IMS stations. Its magnitude was about 3.5, which according to Table 2-2 would correspond to about 100 tons tamped in hard rock, but corresponds to only a few tons yield for an explosion in water. At the IMS and non-IMS stations which recorded regional signals, correlation analysis permits clear seismic detection of the first and much smaller explosion, with yield about

---

[26] This organization, with headquarters in the United Kingdom, has served research communities for decades with authoritative information on seismic source locations, estimated on the basis of detections reported voluntarily by hundreds of institutions operating seismic networks at the national or regional or local scale.

[27] A kilogram is a millionth of a kiloton.

[28] The T-phase travels horizontally from an oceanic source as a hydroacoustic wave in the water, converting to a seismic wave upon incidence at the interface between sea and land. It is then recorded on land as a seismic wave.

a hundred times smaller (i.e., a few tens of kilograms).

The IMS hydroacoustic and seismic networks are complementary for monitoring the world's oceans, in that the hydroacoustic network has highest sensitivity in the southern oceans where IMS seismic networks are weakest, and the seismic networks have high sensitivity in the northern oceans where the IMS hydroacoustic network is weakest.

Monitoring Nuclear Explosions in the Atmosphere

Nuclear tests in the atmosphere are best monitored with radionuclide, infrasound, and electromagnetic systems (including satellite-based optical detection, which is not part of the IMS). The hydroacoustic network also contributes to monitoring for atmospheric explosions above the broad ocean area. In this section we comment on the infrasound and radionuclide IMS networks.

Infrasound Monitoring

Volcanoes, meteorites, explosions, and other natural phenomena and industrial activities produce sound waves in the atmosphere that can be detected at substantial distances from the causative event. This propagation process is modified by wind and temperature variations, even though there is little attenuation at low frequencies (less than 10 Hz). There is a substantial body of knowledge regarding infrasound propagation in the atmosphere, derived in large part from monitoring hundreds of atmospheric nuclear explosions that occurred prior to the Limited Test Ban Treaty of 1963.[29] The international community decided to include an infrasound network as part of the IMS since the CTBTO does not have independent access to satellite-based methods for monitoring atmospheric testing.

The infrasound network of the IMS will consist of 60 stations in 34 countries. Several of these stations are operating as of mid-2002. The sensors are simple, though much effort is going into the development of methods to obtain spatial averaging of local atmospheric pressure variations in order to reduce wind noise. Data are collected at each field site and passed by satellite to the IDC. Because of difficulties in interpreting the arrival time of infrasound signals in terms of distance to the causative source (difficulties due to the effects of wind and temperature on sound velocity in the atmosphere), location estimates of an infrasound sound source are expected to be poor compared to those achieved for seismic events. It can be expected that even a small nuclear explosion in the atmosphere would be associated with signals derived from the capabilities of NTM of some states party to the CTBT.

Figure 2-5 shows the projected 90 percent probability of two station detection thresholds for the IMS infrasound system. It is apparent from Figure 2-5 that thresholds are below 1 kt worldwide and below 500 tons on continents in the northern hemisphere.

Radionuclide Monitoring

Radionuclides provide the most definitive identification of an event as being a nuclear explosion. They offer highly sensitive, albeit slow (multi-day) detection of atmospheric, underwater, and vented underground nuclear explosions. No naturally occurring event that might cause a seismic or acoustic signal of concern will generate a population of radionuclides that

---

[29] Atmospheric testing by France and China, which did not sign the LTBT, continued for several more years.

COLOR PLATES

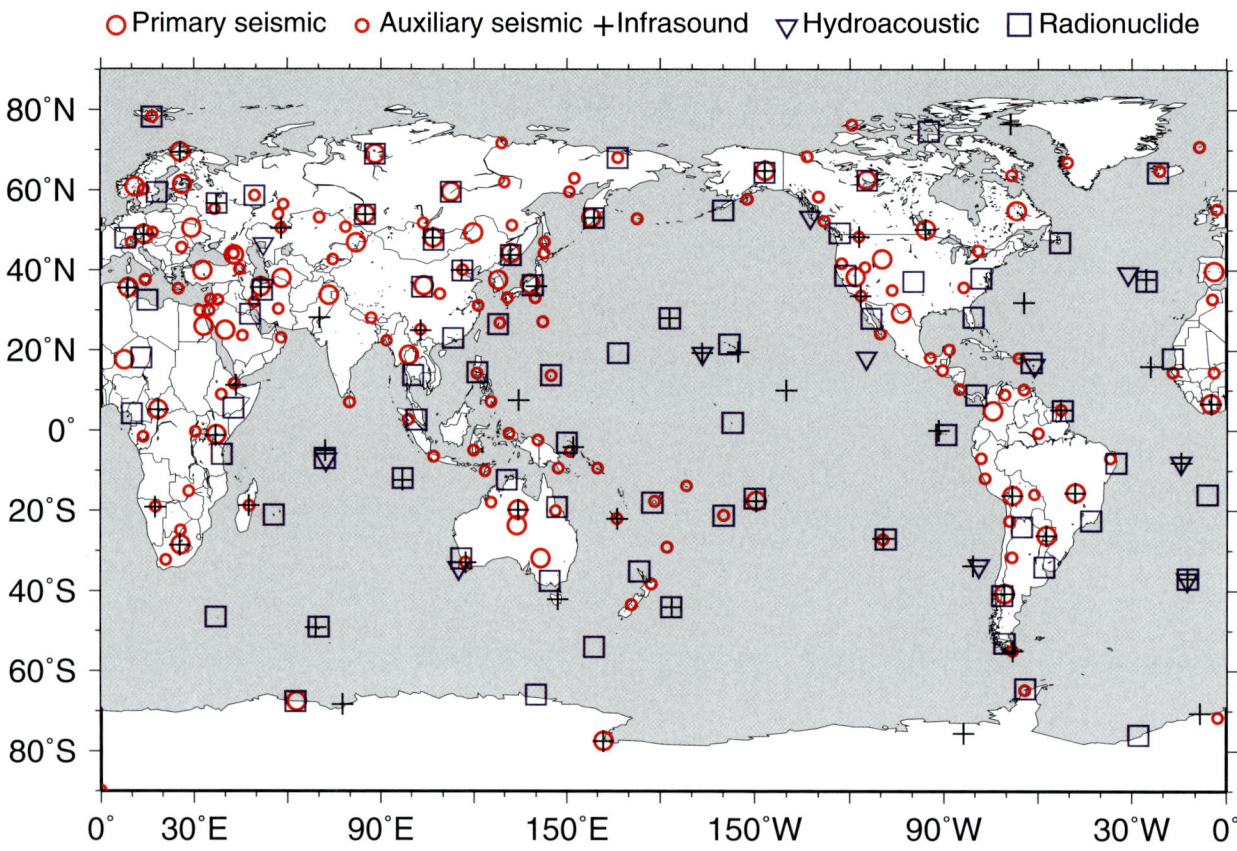

Figure 2-1 A global map showing five networks of IMS stations, which use the technologies of seismology, infrasound, hydroacoustics, and radionuclides. Data from all these stations are telemetered to the IDC in Vienna. Certified laboratories at 16 locations, not shown, contribute to the analysis of radionuclide data. (Figure courtesy of W.Y. Kim.)

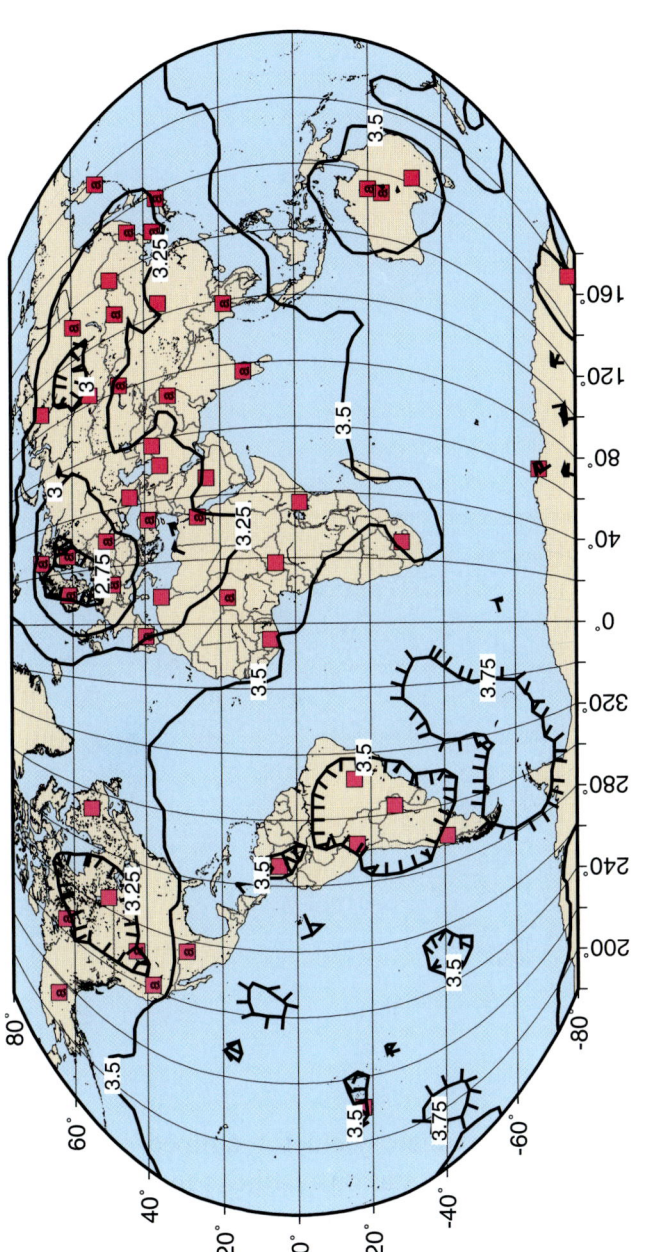

Figure 2-2 Contours of seismic magnitude for which signals would be expected (with signal-to-noise amplitude ratio greater than 3.2, i.e. 10 dB) at three or more stations of the IMS primary seismic network (solid squares), from 90 percent of the events at the contoured magnitude or larger. The contour interval is 0.25 magnitude units. The detection threshold for Europe, Asia, North America, and North Africa is in the magnitude range 3.5 to 3 or lower. (Figure provided by the Center for Monitoring Research)

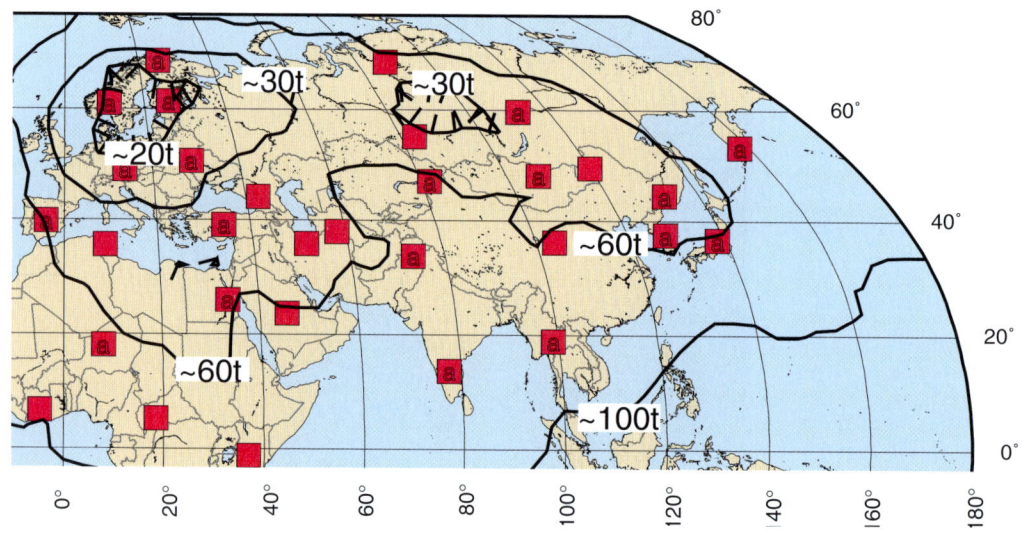

Figure 2-3 Contours of approximate yield for tamped explosions, for which detections can be expected at three IMS primary stations (solid squares). These contours are the same as those of Figure 2-2, but with an expanded view of Europe, Asia and North Africa, and using the approximate yields of Table 2-2 (expressed now in tons rather than kilotons) to interpret seismic magnitudes.

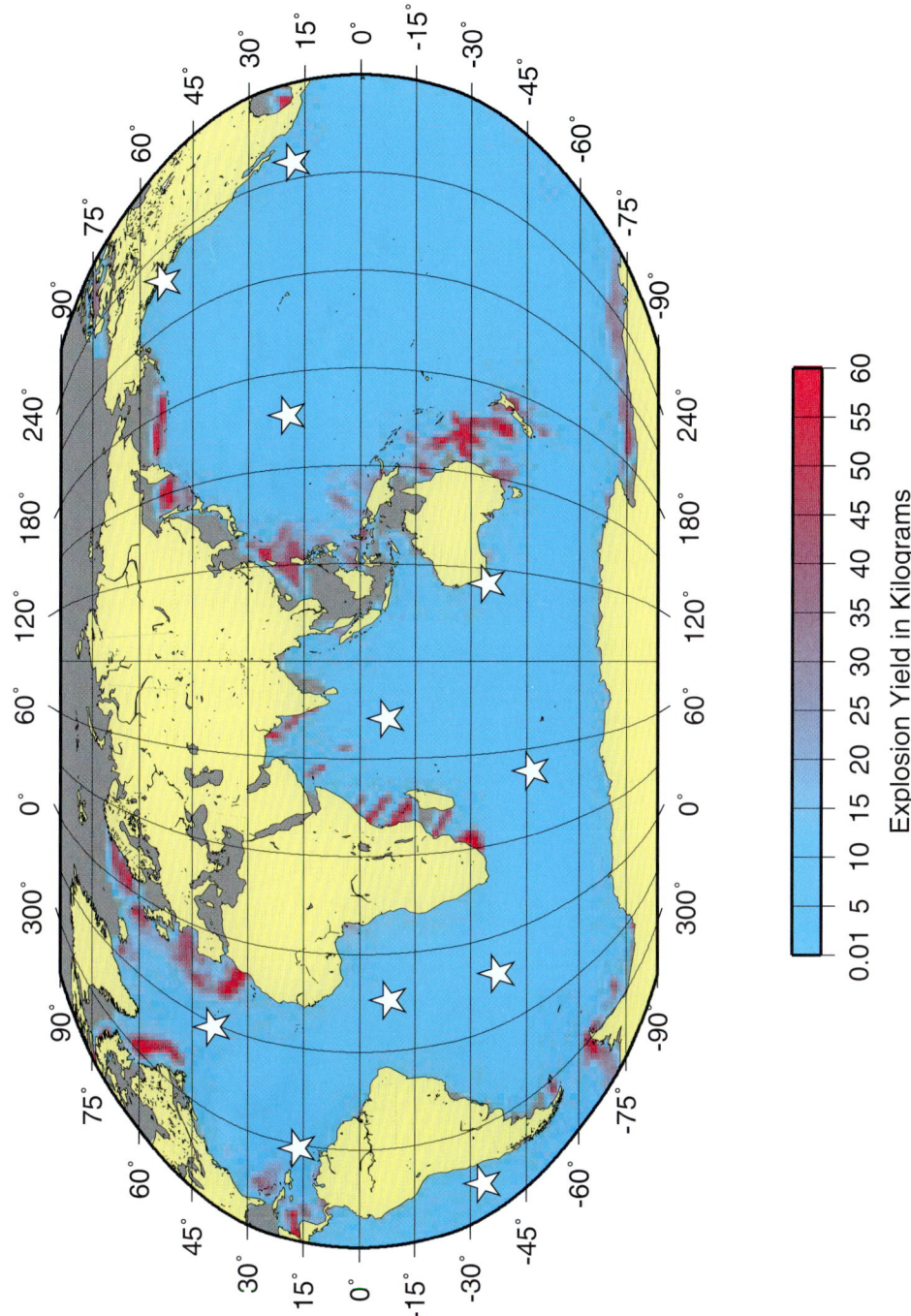

Figure 2-4 Projected 90 percent probable two-station detection thresholds for underwater explosions expressed in kilograms for the IMS network of 11 hydroacoustic stations. (Source: Center for Monitoring Research).

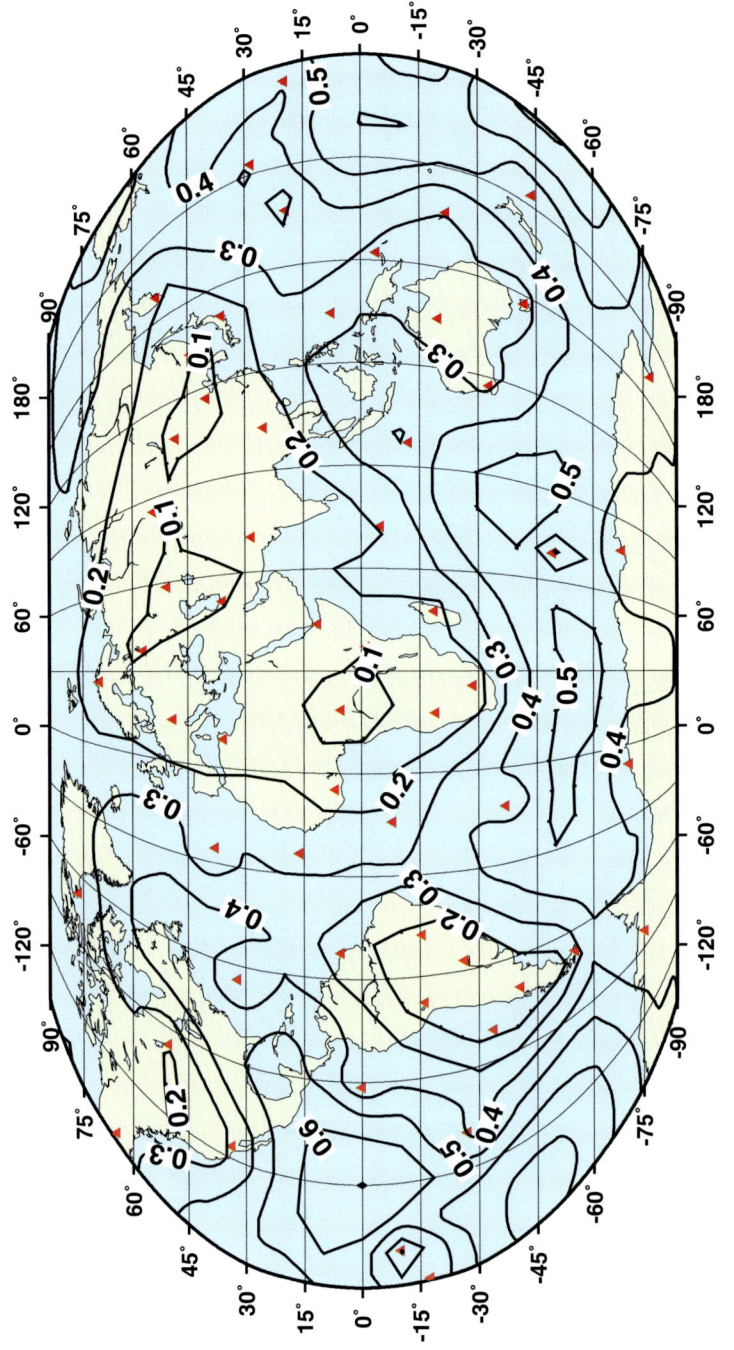

Figure 2-5 Projected 90 percent probable two-station detection thresholds for atmospheric explosions expressed in kilotons for the planned IMS network of 60 infrasound stations. (Source: Center for Monitoring Research.)

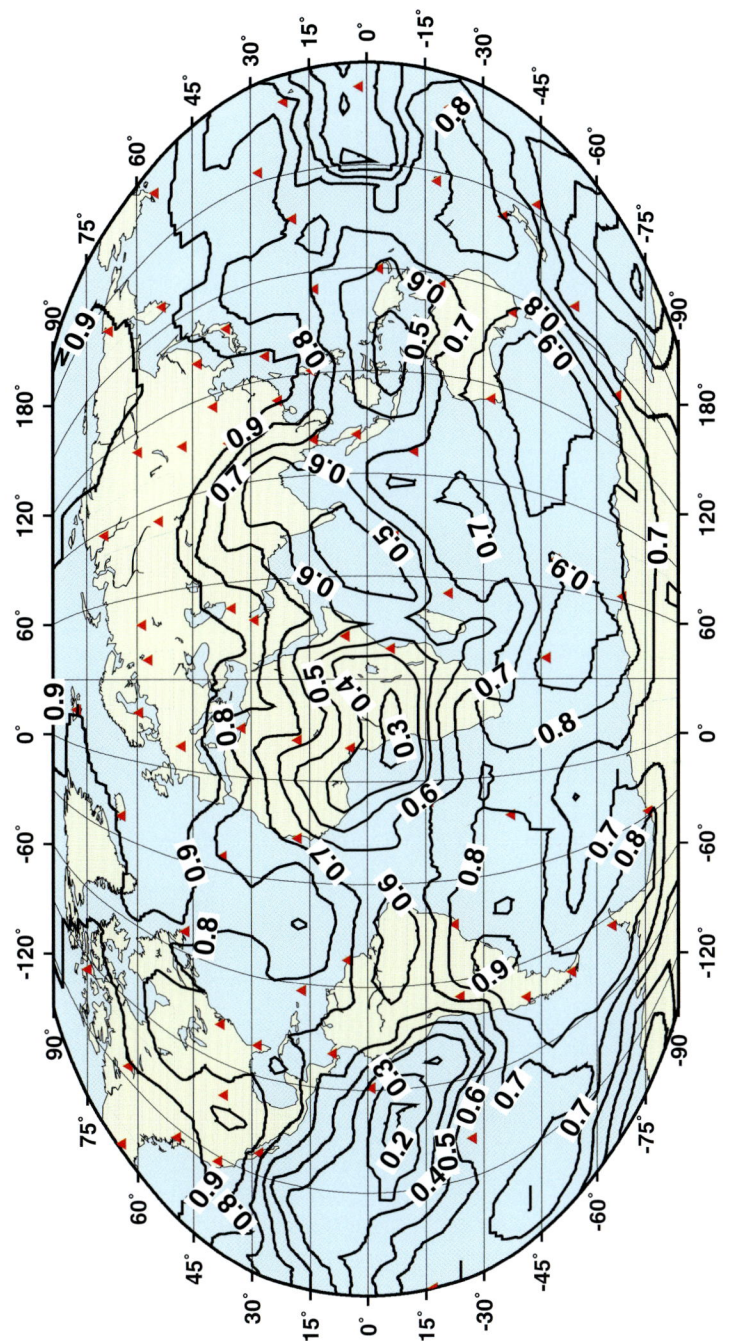

Figure 2-6 Probability of one-station detection of a 1-kiloton atmospheric nuclear explosion within five days by the planned 80-station IMS radionuclide network. (Figure provided by the Center for Monitoring Research.)

would be characteristic of a nuclear explosion. Radionuclide detection has historically been most closely associated with atmospheric nuclear explosions (fallout), but as indicated in Table 2-1, it is an essential element of the IMS for monitoring all environments except for near-space nuclear explosions. The radionuclide element of the IMS comprises a planned network of 80 stations, of which 29 were operational as of January 2002. Capability for noble gas detection (in particular, radioactive isotopes of xenon) is initially required at 40 of these stations.[30] There will also be 16 laboratories capable of performing additional analysis of samples. It appears that this IMS network will have sensitivity on a global scale far better than any radionuclide monitoring system previously available.

All of the 80 stations will have particulate samplers. These pass large volumes of air through a filter that traps particulate matter, concentrating it so that sensitive radiation detection equipment can determine the radionuclides present in the sample. The detection of anomalies in gamma-ray spectra recorded from aerosols sampled from the atmosphere triggers a notification that debris from a nuclear explosion may be present. Capability for automated detection of anomalies in gaseous xenon-isotope abundances will be a particularly sensitive and important diagnostic of a nuclear explosion.[31]

Present estimates of the sensitivity of the radionuclide detection system are 0.1 to 1.0 kilotons for a nuclear explosion on the continents and 1 to 2 kilotons in oceans. Figure 2-6 shows the probability of single-station detection for a 1-kt atmospheric explosion within 5 days. The probability of detection increases with time after the event and exceeds 90 percent nearly worldwide within 10 days. Probabilities of detection exceed 90 percent across most of Europe and Asia and exceed 50 percent over most of the southern oceans.

Even though the radionuclide system can give proof of a nuclear explosion, backtracking the path of detected radionuclides is imprecise and does not provide an accurate estimate of the explosion location.[32] Detection of the characteristic radionuclides of a nuclear explosion would trigger searches for confirming evidence, and information on the location, based on other IMS elements and on NTM.

Monitoring Nuclear Explosions in Space

Nuclear tests in space are best monitored with a variety of electromagnetic sensors, which also have a major role in monitoring explosions in the atmosphere.

Nuclear explosions in space have been suggested by test ban critics as a possible evasion technique since 1958. However, because of the technical complexity, cost, and difficulty of obtaining diagnostic data, the use of space for CTBT evasion appears highly unlikely. Nuclear explosions were conducted in space in the early 1960s by the United States and the Soviet Union to determine some of the effects of such explosions, which were very dramatic—not to learn about the explosive weapons themselves.

Nuclear explosions in space can be detected through a number of signatures: X-rays from explosions of about a kiloton can be detected at ranges exceeding 100 million kilometers,

---

[30] Language in the CTBT Protocol is unusually specific here, stating with respect to 80 radionuclide stations that "All stations shall be capable of monitoring for the presence of relevant particulate matter in the atmosphere. Forty of these stations shall also be capable of monitoring for the presence of relevant noble gases upon the entry into force of this Treaty." The IMS global network of radionuclide sensors has been planned to provide detection if 10 percent radioactive noble gases generated by a 1-kt underground explosion would be vented.

[31] T. Bowyer, et al, "Field Testing of Collection and Measurement of Radioxenon for the Comprehensive Test Ban Treaty," *Journal of Radioanalytical and Nuclear Chemistry* 240 (1999): 109-122.

[32] Though efforts would be made to backtrack the path taken by detected radionuclides, the nature of atmospheric transport leads to great location uncertainties.

comparable to the distance between the Earth and the Sun. Optical effects are observable from space from altitudes between sea level to 100 kilometers over extended ranges. EMP effects are produced through intense currents induced by gamma-rays in the ionosphere. In consequence of these recognized technical monitoring possibilities and the unattractiveness of space testing, a ban on explosions in space was included in the LTBT of 1963.

The CTBT, LTBT, and NPT do not incorporate monitoring of space explosions as part of an international monitoring system. Thus monitoring of space explosions depends on national technical means. Sensors involved in observing space explosions serve the dual role of treaty monitoring and detecting and locating nuclear explosions should they be used in actual combat. Deployment of monitoring equipment continues to depend on the priorities given to the treaty monitoring missions relative to other possible space payloads.

Historically nuclear explosion detection monitoring equipment was deployed from 1963 to 1984 as part of the VELA satellite program. Twelve VELA satellites were launched between 1963 and 1970.[33] Currently nuclear explosion monitoring is a secondary mission assigned to the DSP early warning satellite systems. One sensor on the DSP is provided by the Air Force and several detectors are provided by the Department of Energy. Since 1970, 18 DSP satellites have been so equipped and the lifetime of the detectors has exceeded their design span of five years. Coverage by the DSP satellites has been incomplete. The polar regions of the Earth are not covered and most other areas of the Earth are monitored by only one satellite at any one time. In 1975 nuclear-explosion monitors were added to the GPS navigational system. Currently such detectors are flown on 33 GPS satellites. Because of the orbits of the GPS satellites, the detectors fly in a rather harsh radiation environment. Current payloads include optical detectors (bhangmeters), neutron detectors, five X-ray detectors, and two gamma ray sensors.

Over the next decade an extensive replacement program has been proposed for the sensors to be carried by the follow-on satellite system planned to replace the current satellites beginning in the year 2006. This program includes:
- enhanced optical sensors (providing five times greater sensitivity than the current system)
- autonomous EMP sensors incorporating improved background discrimination
- gamma-ray and neutron detectors
- infrared sensors
- x-ray sensors
- on-board data processing

Which and how many of these improvements will be incorporated in the future is still a matter of extensive discussion because of the conflicting budgetary and other priorities assigned to the DSP and GPS payloads, and those of still other satellites.

Other space-based detectors are also deployed, some of which provide limited additional nuclear-explosion detection capability, and some of which provide technical signatures of nuclear weapon-related activities. Extensive and successful efforts are made to track all objects launched into space. All objects in orbit around the Earth that are large enough to have any role in execution of a nuclear test are also tracked.

To summarize, monitoring of space-based explosions and of atmospheric explosions from space is based on powerful technologies which, provided they continue to be deployed and maintained as needed for this mission, will make it extremely unlikely that such testing can be

---

[33] In 1979, signals received from a VELA satellite indicated a possible clandestine test of a nuclear weapon in the South Atlantic or southwest Indian Ocean. Disagreement persists on whether or not such a test occurred, although the most recent review available (1994) concludes that it did not. The disagreement is due in part to limitations of the instrumentation available at that time. Today's instrumentation for detecting nuclear weapons tests in the atmosphere is sufficiently improved to enable unambiguous interpretation of such events.

conducted without being detected and identified. Attribution might require additional information, such as from use of NTM to interpret missile-launch activities.

The Role of Confidence-Building Measures and On-Site Inspections

Confidence-building measures have already been mentioned in the context of potential problems with mine blasting. The CTBT also specifies that confidence-building measures assist in calibration of IMS stations; and that each State Party undertakes to cooperate with the CTBTO and other states party to the treaty, in carrying out chemical calibration explosions or to provide relevant information on chemical explosions planned for other purposes. A number of information exchanges (between the United States and Russia) and chemical calibration explosions (in Russia and Kazakhstan) have already been carried out, contributing to IMS station calibration.

The CTBT provides for challenge on-site inspections to clarify whether or not an underground nuclear explosion has occurred. Any party to the treaty can request an inspection based on objective information from the IMS and/or National Technical Means (e.g., reconnaissance satellite photographs). The request must give approximate coordinates of the alleged event and proposed boundaries of the area to be inspected (see discussion of event location above). Authorization of an on-site inspection, which is governed by a fast-moving schedule in order to seek evidence that may diminish with time, requires a positive vote by at least 30 of the 51-member Executive Council of the CTBTO.

Inspectors can carry out a range of activities, including measurement of: aftershocks (to help distinguish between earthquake and explosion, and to increase the precision of location); radioactive noble gases or debris (such as argon-37 and krypton-85), which often escape through small cracks following an underground nuclear explosion; surface and vegetation changes due to spallation;[34] and human artifacts characteristic of test activity. While it is impossible to quantify the likelihood that any of these techniques would succeed, a violator would not be able to anticipate how to conceal all potential evidence. If the inspection determined the precise location of a suspect explosion (e.g., by aftershocks), drilling could be undertaken, which if it reached the explosion cavity would be able to obtain an unambiguous sample of radionuclides.

The question has been raised whether the Executive Council would ever give 30 votes to approve an on-site inspection, given the high political stakes involved. It is our opinion that, if a strong case were presented in relation to a "state of concern," an inspection would be approved. Gaining approval for an inspection of a Nuclear-Weapon State or other major powers could present a problem. States that have not conducted a test would presumably welcome the opportunity to clear themselves, if they had not already done so by the preliminary consultations for which provision is made in the treaty.

The more difficult question is whether any country that had actually tested clandestinely would ever permit an inspection. However, refusal in the face of a strong case of possible violation would probably be generally accepted as tacit admission of a violation. In this case, as in the case of unambiguous proof of a violation, the CTBT Conference of States Parties or the Executive Council can bring the issue to the attention of the United Nations for action.

Finally, the question arises as to whether there will be so many unidentified (hence potentially suspicious) small events at the threshold of detection that the system will be overwhelmed with challenges. In reality, given the cost of inspections and the financial and other penalties for

---

[34] Spallation of the ground surface above an underground explosion occurs when the waves generated by the explosion cause surface materials to be thrown upward and then fall back under the influence of gravity.

"frivolous or abusive" requests, as specified in Article IV of the Treaty, it is unlikely that there will be many challenges without strong objective evidence.

In summary, the right of on-site inspection provided by the CTBT constitutes a deterrent to treaty violation whether or not the inspection actually takes place, and it provides a mechanism for the innocent to clear the record.

Research and Development in Support of CTBT Monitoring

Following the first CTBT negotiations, the need to develop a scientific basis for monitoring nuclear testing in all environments was explicitly recognized in 1959. This recognition led to programs of basic and applied research that continue to this day. Monitoring capability has steadily improved as global networks were installed beginning in the 1960s, and as instrumentation and methods of analysis have continued to improve. Safeguard D states that "The United States shall continue its comprehensive research and development program to improve its capabilities and operations for monitoring the Treaty."[35]

The four different monitoring technologies now used by the IMS can all be expected to continue to improve over the coming years.

- Hydroacoustics is well advanced because of work on detection of submarines, but there has been less research on transient signals and still less on the use of hydroacoustic signals to monitor underground and atmospheric explosions. Work is needed in source excitation theory for diverse ocean environments, particularly for earthquakes, and for acoustic sources in shallow coastal waters and low altitude environments. A quantitative understanding of ocean-land and land-ocean coupling would assist in the interpretation of T-phases.

- Infrasound was a recognized research field in the 1950s and up to the early 1970s, but the reduction in atmospheric testing has led almost to the elimination of infrasound studies in the United States over the last 20 years. The IMS system will establish a global array of infrasound sensors to enable routine monitoring of low-frequency sound waves on a global basis for the first time in decades. Research issues associated with CTBT monitoring involve first-order questions about the background noise, involving wind noise reduction and the nature and frequency of events such as volcanic explosions, meteor impacts, and other natural sources. Practical questions arise on the infrasound detectability of mine blasts. Basic information on U.S. monitoring experience has been released and is systematically helping to enhance monitoring capability.[36] But building up the necessary infrastructure to enable effective operation of the IMS network will take several years of support.

- The IMS radionuclide network is far more sensitive and spatially comprehensive than previous networks have been, but its infrastructure can be improved in order to support more effective operations. In addition, there is an opportunity to involve the vigorous research community in atmospheric chemistry, which has for the most part not been engaged in the technical work of improving treaty monitoring. There is considerable potential for mutually reinforcing efforts on the part of the basic-science and treaty-monitoring communities.

- Seismic research on CTBT monitoring is largely engaged in the interpretation of regional waves, with two goals: developing practical methods to improve estimates of event loca-

---

[35] White House, Office of the President, September 22, 1997.
[36] Some very interesting infrasound signals have been received over the years, for example from incoming bolides, which on average impact the atmosphere once a month with energy of a kiloton or more.

tion based on detection of regional waves at IMS stations, and improving methods of event identification using differences between the regional waves excited by earthquakes and explosions. Steady improvements in knowledge of Earth structure can be expected to result in improvements in the accuracy of event location. The development of methods for interpreting seismic waveforms will continue to improve methods of determining the depth of earthquakes, most of which are deeper than 10 km. Events determined to be at this depth, or greater, can only be earthquakes.

Conclusions on CTBT Monitoring Capability

Detection, identification, and attribution of nuclear explosions rest on a combination of methods, some being deployed under the International Monitoring System established under the CTBT, some deployed as National Technical Means, and some relying on other methods of intelligence collection together with openly-available data not originally acquired for treaty monitoring. The following conclusions presume that all of the elements of the IMS are deployed and supported at a level that ensures their full capability, functionality, and continuity of operation into the future.

In the absence of special efforts at evasion, nuclear explosions with a yield of one kiloton or more can be detected and identified with high confidence in all environments. Specific capabilities in different environments are as follows:

- Underground explosions can be reliably detected and can be identified as explosions, using IMS data, down to a yield of 0.1 kilotons (100 tons) in hard rock if conducted anywhere in Europe, Asia, North Africa, and North America. In some locations of interest such as Novaya Zemlya, this capability extends down to 0.01 kiloton (10 tons) or less. Depending on the medium in which the identified explosion occurs, its actual yield could vary from the hard rock value over a range given by multiplying or dividing by a factor of about 10, corresponding respectively to the extremes represented by a test in deep unconsolidated dry sediments (very poor coupling) and a test in a water-saturated environment (excellent coupling). Positive identification as a *nuclear* explosion, for testing less than a few kilotons, could require on-site inspection unless there is detectable venting of radionuclides. Attribution would likely be unambiguous.
- Atmospheric explosions can be detected and identified as nuclear, using IMS data, with high confidence above 500 tons on continents in the northern hemisphere and above one kiloton worldwide, and possibly at much lower yields for many sub-regions. While attribution could be difficult to determine based on IMS data alone, evaluation of other information (including that obtained by NTM) could provide an unambiguous determination.
- Underwater explosions in the ocean can be reliably detected and identified as explosions, using IMS data, at yields down to 0.001 kiloton (1 ton) or even lower. Positive identification as a nuclear explosion could require debris collection. Attribution might be difficult to establish unless additional information was available, as it might be, from NTM.
- Explosions in the upper atmosphere and near space can be detected and identified as nuclear, with suitable instrumentation, with great confidence for yields above about a kiloton to distances up to about 100 million kilometers from Earth. (This capability is based on the assumption that relevant instruments that have been proposed for deployment on the follow-on system for the DSP satellites will in fact be funded and installed.) Such evasion scenarios are costly and technically difficult to implement. If they materialize, attribution will probably have to rely upon NTM, including interpretation of missile launch activities.

The capabilities to detect and identify nuclear explosions without special efforts at evasion are considerably better than the "one kiloton worldwide" characterization that has often been stated for the IMS.[37] If deemed necessary, these capabilities could be further improved by increasing the number of stations in networks whose data streams are continuously searched for signals.

In the history of discussions of the merits of a CTBT, a number of scenarios have been mentioned under which parties seeking to test clandestinely might be able to evade detection, identification, or attribution. With the exception of the use of underground cavities to decouple explosions from the surrounding geologic media and thereby reduce the seismic signal that is generated, none of these scenarios for evading detection and/or attribution has been explored experimentally. And the only one that would have a good chance of working without prior experimentation is masking a nuclear test with a large chemical explosion nearby in an underground mine. The experimentation needed to explore other approaches to evasion would be highly uncertain of success, costly, and likely in itself to be detected.

Thus, the only evasion scenarios that need to be taken seriously at this time are cavity decoupling and mine masking. In the case of cavity decoupling, the experimental base is very small, and the signal-reduction ("decoupling") factor of 70 that is often mentioned as a general rule has actually only been achieved in one test of very low yield (about 0.4 kiloton). The practical difficulties of achieving a high decoupling factor—size and depth of the needed cavity and probability of significant venting—increase sharply with increasing yield. And evaders must reckon with the high sensitivity of the global IMS, with the possibility of detection by regional seismic networks operated for scientific purposes, and with the chance that a higher-than-expected yield will lead to detection because their cavity was sized for a smaller one.

As for mine masking, chemical explosions in mines are typically ripple-fired and thus relatively inefficient at generating seismic signals compared to single explosions of the same total yield. For a nuclear explosion that is not also cavity-decoupled to be hidden by a mine explosion of this type, the nuclear yield could not exceed about 10 percent of the aggregate yield of the chemical explosion. A very high-yield, single-fired chemical explosion could mask a nuclear explosion with yield more comparable to the chemical one, but the very rarity of chemical explosions of this nature would draw suspicion to the event. Masking a nuclear yield even as large as a kiloton in a mine would require combining the cavity-decoupling and mine-masking scenarios, adding to the difficulties of cavity decoupling already mentioned.

Taking all factors into account and assuming a fully functional IMS, we judge that an underground nuclear explosion cannot be confidently hidden if its yield is larger than 1 or 2 kilotons.

Evasion scenarios have been suggested that involve the conduct of nuclear tests in the atmosphere or at the ocean surface where the event would be detected and identified but attribution might be difficult. NTM of the United States and other nations might provide attribution, without being predictable by the evader.

The task of monitoring is eased (and the difficulty of cheating magnified), finally, by the circumstance that most of the purposes of nuclear testing—and particularly exploring nuclear-weapon physics or developing new weapons—would require not one test but many. (An exception would be the situation in which an aspiring nuclear-weapon state had been provided the blueprints for a weapon by a country with greater nuclear-weapon capabilities, and might need only a single test to confirm that it had successfully followed the blueprints.) Having to conduct multiple tests greatly increases the chance of detection by any and all of the measures in use, from the IMS, to national technical means, to sensors in use for other purposes.

---

[37] See, for example, the United Nations Conference on Disarmament Working Paper CD/NTB/WP.225 (Geneva, 1995).

It can be expected, in future decades, that monitoring capabilities will significantly improve beyond those described here, as instrumentation, communications, and methods of analysis improve, as data archives expand and experience increases, and as the limited regions associated with serious evasion scenarios become the subject of close attention and better understanding.

3

Potential Impact of Foreign Testing: U.S. Security Interests and Concerns

This chapter addresses the potential impact on U.S. national-security interests and concerns of the degree of foreign nuclear testing that could plausibly occur without detection under a CTBT regime, or, alternatively, through overt testing. Our principal focus here is on the technical question of what additions to their nuclear-weapon capabilities other countries could achieve through nuclear testing at yields that might escape detection, but we give some attention as well to the related military and political question of the impact of such additions on the security interests and freedom of action of the United States. These questions are embedded in a wider set of political, military, and diplomatic circumstances which, although not in our charge to analyze here, must be mentioned by way of context for the narrower questions we address.

Currently the United States is the preeminent nation in the world, measured in political, economic, and military terms. In the military dimension, the United States possesses dominant conventional forces as well as deployed and reserve nuclear weapons of mature and amply tested design. Should nuclear weapons proliferate widely across the globe, U.S. military pre-eminence will be diminished. Nuclear weapons are the "great equalizer" among the world's strong and weak military powers. The freedom of action of the United States in exploiting its conventional military superiority will be limited if nations not now possessing nuclear weapons acquire them.

The primary diplomatic tool for restraining the proliferation of nuclear weapons has been the Nuclear Non-Proliferation Treaty (NPT), which entered into force in 1970. That treaty divided its signatories into Nuclear-Weapon States (NWS) and Non-Nuclear-Weapon States (NNWS), where the Nuclear-Weapon States are those nations that had manufactured and exploded a nuclear weapon prior to January 1, 1967. Ever since the enactment of the NPT, achievement of a CTBT has been a litmus test of the willingness of the Nuclear-Weapon States to meet their obligations under Article 6 of the NPT. Nonetheless, most states would probably continue to adhere to the NPT without a CTBT, and therefore could neither acquire nuclear weapons nor test them. But the absence of a CTBT limiting the five Nuclear-Weapon States increases the possibility that some might leave the NPT in order to test—thereby creating a dynamic of proliferation and competition.

By giving up their highly visible right to testing, the Nuclear-Weapon States were seen to be consenting to a halt to the modernization and diversification of their nuclear arsenals, thus at least plausibly beginning a process of de-emphasizing the role of nuclear weapons in international relations. The linkage of enactment of a CTBT to the future viability of a non-proliferation regime was explicitly recognized both in the preamble to the NPT and in the 1995 Review and Extension Conference, which converted the NPT into a treaty of indefinite duration.

The potential impact on U.S. security interests and concerns of the foreign nuclear tests that could plausibly occur without detection in a CTBT regime can only be meaningfully assessed by comparison with two alternative situations—the situation in the absence of a CTBT, and the situation in which a CTBT is being strictly observed by all parties. The key questions are: How much of the benefit of a strictly observed CTBT is lost if some countries test clandestinely within the limits imposed by the capabilities of the monitoring system? In what respects is the case of limited clandestine testing under a CTBT better for U.S. security interests—and in what respects worse—than the case of having no CTBT at all? If some nations do not adhere to a CTBT and test openly, how do the technical and political impacts differ from a no-CTBT era?

In these comparisons, two kinds of effects of nuclear testing by others on U.S. security interests and concerns need to be recognized: the *direct* effects on the actual nuclear-weapon capabilities and deployments of the nations that test, with implications for military balances, U.S. freedom of action, and the possibilities of nuclear-weapon use; and the *indirect* effects of nuclear testing by some states on the aspirations and decisions of other states about acquiring and deploying nuclear weapons, or about acquiring and deploying non-nuclear forces intended to offset the nuclear weapons of others. A CTBT, to the extent it is observed, brings security benefits to the United States in both categories—limitations on the nuclear-weapon capabilities that others can achieve, and elimination of the inducement of states to react to the testing of others with testing and/or deployments of their own.

A nuclear test or series of tests affects the nuclear-weapon-related capabilities of the state that tests—and, if detected by other countries, may affect their aspirations and decisions relating to nuclear weapons—but whether nuclear testing actually leads to weaponization and, beyond that, to deployment, depends on additional factors. These may include a country's motivation to acquire nuclear weapons, as well as its production of or access to plutonium or enriched uranium for fission weapons, the necessary tritium for boosted fission weapons and boosted primaries for thermonuclear weapons, and the lithium-deuteride "salt" that is used in thermonuclear weapons.

Another important factor is the means of delivery, some of which impose greater demands on the nuclear-weapon payload. Although many people appear to believe that the threat from newly nuclear countries is dependent on their possession of an intercontinental ballistic missile (ICBM), even the 1998 Rumsfeld Commission charged with evaluating such ICBM threats called attention very emphatically to the availability of other means of delivery that would accommodate larger, heavier warheads.[1] The possibilities include delivery by military and civilian aircraft, by short-range ballistic missiles and cruise missiles that could be launched from ships near U.S. shores, by truck or car after a weapon has been smuggled across U.S. borders, or by a ship entering a U.S. port with a nuclear weapon on board.

The factors beyond nuclear testing that affect weaponization and deployment—and thus affect the actual military threats that new nuclear-weapon capabilities can pose to the United States—are far beyond the mandate and capacity of this committee to address. Analysis of these factors is the daily meat of intelligence assessment and, while a necessary part of a "net assessment" of threats to U.S. security interests, cannot practically be incorporated into our treatment of the implications of potential clandestine nuclear testing under a CTBT. We confine ourselves here to the nuclear-weapon potentials likely to be achievable with nuclear testing in various yield

---

[1] D.H. Rumsfeld, et al., *Executive Summary of the Report of the Commission to Assess the Ballistic Missile Threat to the United States* (July 15, 1998).

ranges (as well as without testing at all), referring to such factors as delivery systems only in the context of their interaction with the question of what kinds of nuclear weapons can be developed.

In the remainder of this chapter, we first consider the characteristics of the two reference cases—no CTBT and a CTBT scrupulously observed—in relation to the kinds of advances in the nuclear-weapon capabilities of other countries that could be expected and how these two situations would affect U.S. security interests and concerns. We then discuss the advances that could plausibly be made by clandestine testing in various yield ranges, under a CTBT, by countries with greater prior nuclear test experience and/or design sophistication and those with lesser experience and/or sophistication.

Following this general comparison of the clandestine-testing case with the two reference cases, we offer some additional observations, on a country-by-country basis, for particular states: Russia, China, Pakistan, Iraq, Iran, and North Korea. The first two in this group are Nuclear-Weapon States that have possessed nuclear weapons for 50+ and 35+ years, respectively, and that have long been able to deliver these weapons against the United States with ICBMs as well as by other means. Given the capabilities they have long possessed, further improvements in their nuclear weapons would be of limited security impact on this country. India and Pakistan have carried out limited nuclear testing, and weaponization and deployment plans appear to be hanging in the balance. Iraq, Iran, and North Korea have manifested nuclear ambitions in recent years, but have conducted no tests.

Two Reference Cases: No CTBT and the CTBT Strictly Observed

No CTBT

In the reference case of no CTBT at all, the Nuclear-Weapon States Party to the NPT would be able to test without legal constraint in the underground environment (except for the 150-kiloton limit agreed to by the United States and Russia under the bilateral Threshold Test Ban Treaty), and non-parties to the NPT would similarly be able to test without legal constraint. Non-Nuclear-Weapon States Party to the NPT would be legally constrained from testing.

As discussed further in the country-specific treatments in this chapter, all of the countries that would be free to test in the absence of a CTBT have some motivation to do so. (This includes the United States, which if unconstrained might test to explore further improvements in the safety of its nuclear weapons, to test nuclear weapons effects, to explore new nuclear-weapon concepts, and, occasionally, to add to confidence in solutions devised in the Stockpile Stewardship Program for age-related defects in stockpiled weapons.) Given that the United States has already conducted more than 1,000 nuclear tests, however, compared with 715 of the Soviet Union, 215 of France, 45 of Britain, and 45 of China, and given the relative maturity of U.S. designs, it is likely that the other countries that would be unconstrained in the absence of a CTBT could make more relative progress with additional tests than could the United States. India and Pakistan claim six nuclear tests each.

China and Russia might use the option of testing to make certain refinements in their nuclear arsenals, which are discussed further in the country-by-country treatment. In the case of Russia, it is difficult to envision how such refinements could significantly increase the threats to U.S. security interests that Russia can pose with the previously tested nuclear-weapon types it already possesses. In the case of China, further nuclear testing might enable reductions in the size and weight of its nuclear warheads, as well as improved yield-to-weight ratios. Such improvements would make it easier for China to expand and add multiple independently targetable re-entry vehicles (MIRVs) to its strategic nuclear arsenal if it wanted to do so, and changes in these directions would affect U.S. security interests. But China could also achieve some kinds of improvements in its nuclear weapons without nuclear testing, and if it wanted to do so it could achieve considerable expansion and MIRVing of its arsenal using nuclear-weapon types it has already tested.[2]

More importantly for U.S. security interests and concerns, India and Pakistan could use their option of testing, as non-parties to the NPT, to perfect boosted fission weapons and thermonuclear weapons. This would greatly amplify the destructive power available from a given quantity of fissile material and the destructive power deliverable by a given force of aircraft or missiles. (Of course they might also do this under a CTBT that they had not signed, but the absence of a CTBT and the resumption of testing by others would make it politically much easier for them to do so.) The likelihood that either of these countries would use nuclear weapons against the United States seems very low, but the United States and its allies would nonetheless have serious concerns about the increase in nuclear-weapon dangers and arms-race potential in and around South Asia that such developments would portend.

Plausibly larger than the direct effects of testing by Nuclear-Weapon States and non-parties to the NPT in the absence of a CTBT is the potential indirect effect of a resumption of such testing in the form of a breakdown of the NPT regime, manifested in more widespread testing (by such countries as North Korea, Iraq, and Iran, for example), which could lead in turn to nuclear weapons acquisition by Japan, South Korea, and many others. With sufficient testing, many countries would be able to master boosted fission weapons and thermonuclear weapons. Some might do this, in a world of more and more nuclear-armed states, not to solve real security problems but simply for reasons of prestige or "equity."

A future no-CTBT world, then, could be a more dangerous world than today's, for the United States and for others. In particular, the directions from which nuclear attack on the United States and its allies would have become conceivable—and the means by which such attack might be carried out (meaning not only ICBMs but also, among others, ship-based cruise missiles, civilian as well as military aircraft, and truck bombs following smuggling of the weapons across U.S. borders)—would have multiplied alarmingly.

We note, finally, that while a CTBT does not add to the obligations of Non-Nuclear-Weapon State NPT parties not to conduct nuclear tests or to acquire nuclear weapons, it does greatly strengthen the capacity of the international community to monitor nuclear testing (both by internationally agreed remote detection of nuclear tests in any country and, in the case of parties to the CTBT, by the rights to on-site inspections). These contributions of a CTBT to international monitoring capabilities would be absent in the no-CTBT scenario.

---

[2] "China has had the technical capability to develop multiple RV payloads for 20 years. If China needed a multiple-RV (MRV) capability in the near term, Beijing could use a DF-31-type RV to develop and deploy a simple MRV or multiple independently targetable reentry vehicle (MIRV) for the CSS-4 in a few years. MIRVing a future mobile missile would be many years off." National Intelligence Council, *Foreign Missile Developments and the Ballistic Missile Threat to the United States Through 2015* (September 1999, unclassified).

CTBT Strictly Observed

Even scrupulous observance of a CTBT would not preclude the emergence of additional nuclear-armed states. Most of the activities involved in the development of nuclear weapons do not involve nuclear testing. The mature nuclear-weapon states are widely credited with sophisticated computational techniques, advanced hydrodiagnostic methods, and extensive knowledge of relevant materials properties. Similar, even if less-advanced capabilities available to would-be nuclear states would allow constructing some relatively simple types of nuclear weapons that could be expected to work (albeit with far less efficient use of nuclear material and much lower yield-to-weight ratios than in designs attainable with the help of nuclear testing).

Computational capabilities have evolved from the three multiplications per second of the 1944 card-punch calculator, to the ten million operations per second of the CDC 7600 mainframes of the 1970s and early 1980s, on which much of the U.S. enduring stockpile was designed, to the billion operations per second of the personal computer in the year 2000. Unclassified computer programs for one- and two-dimensional hydrodynamics and neutronics calculations are widely available. Neutron cross-section libraries and much relevant equation-of-state data are in the public domain. High-explosive and detonator technologies are widely dispersed. Optical techniques using high-speed framing and streak cameras are routine state of the art in industry.

So-called subcritical tests for the study of the properties of fissile materials subject to shock from high explosives avoid the initiation of a nuclear chain reaction and are not prohibited by the CTBT. They typically involve small plates or other shapes formed of plutonium or uranium. Hydrodynamic tests involve material in weapon configuration but are arranged to avoid criticality, either by means of reducing the scale below critical mass or by replacing the plutonium or U-235 with a simulant material. These, too, are not prohibited by the CTBT. The "pin" technique basic to the design of implosion systems is 55 years old, modified only by the use of modern fast-recording systems. (The "pins" are fine wires or optical fibers that report time of contact with an imploding metal shell.)

Relatively unobtrusively, these tools can be used to establish a nuclear-weapon design capability. Dynamic radiography capabilities, which were not available for the early 1960s U.S. stockpile designs, are somewhat more difficult to conceal. But all these development tools, short of an actual nuclear proof test, would be available under the CTBT. They can be used by experienced nuclear-weapon states to refine their understanding and by would-be proliferators to develop simple nuclear-weapon designs in which it would be possible to have some confidence without testing them (although such conduct by a Non-Nuclear-Weapon State Party to the NPT would violate that treaty).

In 1945, after all, Hiroshima was devastated by a nuclear weapon that had never been subject to a nuclear-explosion test. This weapon, containing weapon-grade uranium, weighed some 8,000 pounds and had a yield of about 13 kilotons. In such a weapon, a U-235 projectile is fired into a U-235 sleeve, to form a compact configuration that exceeds a critical mass. For any nation with a modest technical competence, laboratory measurements would suffice for such a uranium-235 gun design, together with firing the gun with a dummy projectile. Such developments could take place without violating the CTBT.

Without nuclear testing, South Africa produced six modernized, lighter U-235 gun-type weapons, which were dismantled when South Africa joined the Non-Proliferation Treaty as a Non-Nuclear-Weapon State. Knowledge of these and of the fact that the United States once possessed large numbers of artillery-fired gun-type nuclear shells might lead a proliferant country to a system much lighter and smaller than the Hiroshima weapon.

Access to highly enriched uranium, either by indigenous production or by purchase or theft abroad, is key to the acquisition of gun-type weapons. Indigenous production is a substantial effort, subject to detection; acquisition abroad might also be discovered. Of course, acquisition of U-235 for weapons purposes by a Non-Nuclear-Weapon State belonging to the NPT would violate that treaty.

U-235 can also be used in implosion-type weapons, which require considerably less of this material than do gun-type devices. This approach, then, increases the number of weapons that can be made from a given stock of U-235. A country possessing U-235 might develop without testing an implosion weapon that, although large and heavy by virtue of the quantity of high explosive used, could readily be delivered by ship, commercial aircraft, truck, or train.

Nagasaki was destroyed by an implosion weapon containing about 6 kg of plutonium. It weighed 9,000 pounds and had an explosive yield of about 20 kilotons. Fifty-five years later, and with all the information that has since been declassified, a state with the requisite technical skills in explosives, electronics, and metallurgy could with some confidence reproduce the Nagasaki device without the full-scale test the United States conducted in New Mexico on July 16, 1945. Many non-nuclear tests would be needed to demonstrate the mastery of the technology, and there would be some uncertainty in yield. A weapon weighing 1,000-2,000 pounds might similarly be built, with somewhat less confidence; this might resemble the U.S. Mark-7 bomb of 1951 that weighed 1,800 pounds.

The task of perfecting an implosion weapon is more difficult than the path leading to a U-235 gun-type weapon, but is essential if plutonium is to be used and also provides, as noted above, a path to a weapon using less U-235 than a gun design requires. Technology transfer—authorized or unauthorized, and ranging from tips about dead-end or productive approaches, to transfer of computer codes, to precise working drawings and specifications, to actual transfer of nuclear explosive devices—could greatly ease a recipient state's path to relatively light and compact implosion weapons and could reduce the number of nuclear tests needed to master these. A single full-yield test would validate both the legitimacy of a blueprint and success in reproducing the object, but that test might be of yield too high to be concealed. Access to plutonium for an implosion weapon, moreover, would require either indigenous production in a nuclear reactor or acquisition from outside sources. Either acquisition or clandestine reprocessing of plutonium from nuclear reactors incurs risk of detection.

The size and weight of fission bombs that could be developed confidently without nuclear testing limit the available means of delivery. Transport aircraft, ships, trucks, and trains can carry any nuclear weapon. The most common missile of 300-kilometer range, the SCUD, has a payload capacity of 1,000 kg. The extended-range SCUD used by Iraq against Israel and Saudi Arabia in the Gulf War can carry 500 kg. The 3-stage Taepo Dong-2, under development but as yet untested by North Korea, could deliver a 700-kg payload anywhere in the United States.[3]

In summary, if a CTBT was scrupulously observed, nuclear threats to the United States could still evolve and grow, but the range of possibilities would be considerably constrained. Boosted fission weapons and thermonuclear weapons would be confined to the few countries that already possess them and to those to which such weapons might be transferred, or to which designs might be communicated with sufficient precision that a trusting and competent recipient

---

[3] Testimony by R.D. Walpole, U.S. Senate. 1999. Committee on Foreign Relations. *Ballistic Missiles: Threat and Response.* 106th Congress, 1st session. September 16.

might be able to reproduce them. Other countries might have less stringent confidence requirements than does the United States, but, in general, they also are much more limited in the technology available for pursuing an exact reproduction; substitution of materials or techniques might bring uncertainty or even failure. Perhaps most importantly, in a world in which nuclear testing had been renounced and the NPT remained intact, nuclear proliferation would be opposed by a powerful political norm in which Nuclear-Weapon States and other parties to the NPT and CTBT would find their interests aligned.

Evasive Testing Under a CTBT

In the case we now wish to compare to the no-CTBT and rigorously-observed-CTBT reference cases—that of clandestine testing under a CTBT within the limits imposed by the monitoring system—we distinguish between two classes of potential cheaters, those with greater prior nuclear test experience and/or design sophistication and those with lesser prior experience and/or sophistication. The purposes and plausible achievements for testing at various yields by countries with little versus extensive prior nuclear test experience are summarized in the following table. Table 3-1 (see next page) describes what could be done, not necessarily what will be done. The case of subcritical testing—legal under a CTBT—has been discussed above under the category of the scrupulously observed CTBT. In the following subsections, we elaborate on the other yield categories in this table.

Tests Conducted Underground Without Fear of Detection By Seismic Signals

For purposes of our discussion, "hydronuclear tests" refer to those with a nuclear yield below 0.1 ton of high-explosive equivalent (0.0001 kt). Their primary utility is to conduct so called one-point safety tests.[4] A series of such tests—which are difficult to design and implement for an experienced Nuclear-Weapon State and even more so for states with little or no testing experience—can determine whether a full-scale weapon would provide a tolerably low yield if the high explosive were accidentally detonated at the single most hazardous point. In this yield range, the decompression and disassembly of the plutonium is little affected by the nuclear reaction, and the yield is so low that it gives little information of value for designing full-scale weapons. The U.S. definition of tolerable yield for a one-point detonation is 2 kg of high explosive (HE) equivalent. If there were containment provided up to, say, 10 tons, a clandestine test below that limit would stand little chance of discovery by seismic, infrasound, hydroacoustic, or radionuclide detection schemes; in this situation, an experienced state would need fewer tests to demonstrate one-point safety than if each test were strictly limited to 2 kg. Of course, states newly acquiring nuclear weapons might not be concerned initially about one-point safety at all.

---

[4]For the one-point safety tests authorized by President Eisenhower to be conducted in secrecy during the moratorium he initiated in 1958, a yield limit was set of four pounds (2 kg) of high explosive equivalent. In its historical record (see V. Mikhailov, ed., *Nuclear Testing in the USSR*, vol. 1, VNIIEF: Sarov, 1997, p. 95.), Minatom defines a hydronuclear test as one with a yield less than 100 kg of high explosive equivalent. As for "nuclear test," Minatom adopted the definition developed in the 1990 Protocol to the Threshold Test Ban Treaty—the same one used by the U.S. Department of Energy—in preparing its comprehensive list. A test is defined as either a single explosion, or two or more explosions fired within 0.1 second of one another within a circular area with a diameter of two kilometers. The yield is the aggregate of all of the explosions. The 715 Soviet nuclear tests thus involved 969 explosions; in addition, Russia reports about 90 hydronuclear tests. (See, e.g., the data in Natural Resources Defense Council, "Table of Known Nuclear Tests Worldwide: 1945-69 and 1970-96," at http://www.nrdc.org/nuclear/nudb/datab15.asp.)

Table 3-1 Purposes and Plausible Achievements for Testing at Various Yields

Yield	Countries of lesser prior nuclear test experience and/or design sophistication[5]	Countries of greater prior nuclear test experience and/or design sophistication
Subcritical testing only (permissible under a CTBT)	• Equation-of-state studies • High-explosive lens tests for implosion weapons • Development & certification of simple, bulky, relatively inefficient unboosted fission weapons	same as column to left, plus • limited insights relevant to designs for boosted fission weapons
Hydronuclear testing (yield < 0.1 t TNT, likely to remain undetected under a CTBT)	• one-point safety tests (with difficulty)	• one-point safety tests • validation of design for unboosted fission weapon with yield in 10-ton range
Extremely-low-yield testing (0.1 t < yield <10 t, likely to remain undetected under a CTBT)	• one-point safety tests	• validation of design for unboosted fission weapon with yield in 100-ton range • possible overrun range for one-point safety tests
Very-low-yield testing (10 t < yield < 1-2 kt, concealable in some circumstances under a CTBT)	• limited improvement of efficiency & weight of unboosted fission weapons compared to 1st-generation weapons not needing testing • proof tests of compact weapons with yield up to 1-2 kt (with difficulty)	• proof tests of compact weapons with yield up to 1-2 kt • partial development of primaries for thermonuclear weapons
Low-yield testing (1-2 kt < yield < 20 kt, unlikely to be concealable under a CTBT)	• development of low-yield boosted fission weapons • eventual development & full testing of some primaries & low-yield thermonuclear weapons • proof tests of fission weapons with yield up to 20 kt	• development of low-yield boosted fission weapons • development & full testing of some primaries & low-yield thermonuclear weapons • proof tests of fission weapons with yield up to 20 kt
High-yield testing (yield > 20 kt, not concealable under a CTBT)	• eventual development & full testing of boosted fission weapons & thermonuclear weapons	• development & full testing of new configurations of boosted fission weapons & thermonuclear weapons

A Nuclear-Weapon State could in principle use a series of hydronuclear tests to validate new designs for unboosted nuclear weapons in the yield range of 10 tons but probably not to 1 kiloton. This is difficult, but could be done with appropriate instrumentation. It would require a large quantity of plutonium or enriched uranium, because multiple experiments would need to be done, each with almost the full amount of fissionable material needed for a complete weapon. Advice from an experienced tester could reduce the number of tests required.

Successful Evasion Possible But By No Means Assured

The range from 0.1 ton to 1 kiloton is categorized by the Russian nuclear establishment as that of "very-low-yield tests." For purpose of analysis we break this range into two parts—extremely low yield" from 0.1 to 10 tons, and "very-low-yield" from 10 tons to 1-2 kt. Tests toward the lower end of the extremely-low-yield category would be easy to conceal from seismic monitoring under a CTBT. In the higher part of this category, such tests could serve a country with little prior test experience for the demonstration of one-point safety somewhat more readily than would be the case if there were a firm restraint to avoid exceeding 0.1 tons. A state with

---

[5]That is, lacking an adequate combination of nuclear-test data, advanced instrumentation, and sophisticated analytical techniques, and without having received assistance in the form of transfer of the relevant insights.

experience in testing and design might use tests in this range to develop, with some confidence, weapons with yields up to about 100 tons.

The "very-low-yield" range from 10 tons to 1-2 kt could serve either category of country to develop and validate a deployable weapon of 10-ton to 1 or 2-kt yield. With a series of tests in the 1 to 2-kt range, an inexperienced state might be able to improve the efficiency and yield-to-weight of unboosted fission weapons compared to the performance of the first-generation weapons that could be developed and deployed with some confidence without any testing at all. Concealment of tests in this yield range is plausible under some circumstances, but increasingly difficult as the 1-kiloton level is approached, and much more difficult for inexperienced testers than for experienced ones. Working closely with experienced personnel might permit an inexperienced state to manage with fewer tests. Under some circumstances, such technology transfer could also increase the probability of successful concealment. In the case of experienced Nuclear-Weapon States, tests in this range might serve to help partially develop primaries for thermonuclear weapons.

In our treatment of nuclear-test monitoring above, we conclude that 1 to 2 kt is the practical upper limit of effective decoupling. A test of this yield would provide data helpful for the partial development of a primary for a thermonuclear weapon. But deployment of such an untested component by one of the five Nuclear-Weapon States, which have available fully tested primaries of adequate yield, would not increase the state's capability and would reduce its confidence in its stockpile. A state that has not yet fully tested primaries could not rely on a primary test of less than full yield.

Unlikely to be Concealable

In the "low-yield" range of 1 kt to 20 kt, states with extensive nuclear test experience could develop and fully test primary nuclear explosives and low-yield thermonuclear weapons. Proliferants could do the same. Either could proof test a fission weapon with a yield up to 20 kt, but concealment is highly unlikely. If done openly, such nuclear explosions might have political as well as technical goals. But the political goals would not be achieved by clandestine tests, and clandestine achievement of technical goals would be precluded since tests above 1 to 2 kilotons could not be concealed with confidence.

Impossible to Conceal

The "high-yield" range in excess of 20 kt would normally be used in the absence of a CTBT to test new configurations of boosted fission weapons or thermonuclear weapons. As discussed above in our treatment of nuclear-test monitoring, any nation with a nuclear explosive could detonate it on a barge or small boat on the open ocean. Such a test would likely be detected, identified, and located, but might be attributed only with some difficulty to the nation responsible. A single such test might be attempted by a proliferator—checking the performance of an implosion weapon or even a boosted implosion weapon—as a proof test, before undertaking deployment.

Assessment of the Impact on U.S. Security Interests of Nuclear Weapons Tests of Selected Countries

We first discuss Russia and China, both Nuclear-Weapon States under the NPT, possessing long-standing nuclear-weapon-development programs and previously tested nuclear weapons of a variety of types. We then take up Pakistan and India—non-participants in the NPT with early-stage nuclear weapon programs and very limited test experience. Finally, we discuss North Korea, Iran, and Iraq, all three of which are NPT members but have been involved in nuclear-weapons activities to a varying extent.

Russia and China—States with Mature Nuclear Weapons Programs and Test Experience

Some motivations for evasive testing by Russia and China relating to their existing stockpiles—life-extension programs, safety, and confidence in remanufactured primaries, for example—are no more threatening to U.S. security interests than is an assumed ability of these nations to maintain their stockpiles without testing. For instance, if (as has been suggested) Russia were to employ clandestine nuclear-explosion testing to help maintain the safety and reliability of its stockpile, that would directly impact U.S. security interests only if the United States were unable to maintain its weapons safe and reliable without nuclear testing and thus suffered a comparative disadvantage.

This is not to condone clandestine nuclear testing by anyone; such testing always carries a risk of detection and is therefore dangerous to the non-proliferation regime. And in that way, it is harmful to the security interests of the United States. But potential undetected Russian and Chinese evasive testing is not relevant to the maintenance of U.S. nuclear weaponry. As noted in Chapter 1, we judge that the United States has the technical capabilities to maintain the reliability of its existing stockpile without testing, irrespective of whether Russia or China decides they need to test in order to maintain the reliability of theirs.

<u>Russia</u>

- Without CTBT

Without a CTBT, Russia could have an incentive to test, given the large changes in its military situation compared to that of the Soviet Union. Russia has renounced the Soviet nuclear doctrine of "no first use" of nuclear weaponry, and some members of the Russian nuclear weapons establishment have publicly advocated bolstering the new first-use doctrine by building thousands of new-design tactical nuclear weapons of very low yield—perhaps 10 to 100 tons. In addition, Russian weapon designers have had a long-standing interest in special effects, such as enhanced-radiation weapons, and these might be developed because of their inherent challenge and interest, as seen by the weapon designers, and for battlefield or sub-strategic use, as viewed by some military writers.

Russia might also want to test to reduce the cost of maintaining its nuclear forces. Russian nuclear weapons are remanufactured on a 10-year cycle, a substantial maintenance burden. Probably the nuclear organization would prefer weapons with longer life, with testing to permit some redesign of such weapons and to demonstrate performance. This would likely require yields in excess of 10 kt—such tests are permissible underground in the absence of a CTBT, but not plausibly concealable with a CTBT in force.[6]

Because of the mature state of Russian nuclear weapons technology, unrestrained Russian nuclear testing would not directly impair U.S. security interests; but the indirect effects could be substantial—especially in eroding the non-proliferation regime and legitimizing the acquisition and test of nuclear weapons by other states, in particular those neighboring Russia.

- Russian advances strictly respecting a CTBT

Despite its vast experience in nuclear-weapon design and test, Russia could not confidently develop new nuclear weaponry without violating the zero-yield CTBT. Although it might seem a simple matter to design a weapon of 10 to 100 tons yield, with normal design approaches it would be difficult to have confidence in the yield.

- Evasive testing within the CTBT

As explained in detail in Chapter 2, in no case could a country in Eurasia, including Russia, have high confidence of concealing a test over 1-2 kilotons from seismic detection. Without constructing a large decoupling cavity, the limit to concealed testing is much lower.

With a 1-kt test not concealable at its operating test site, though marginally possible in areas particularly suited for cavity decoupling (e.g., salt domes), Russia could potentially do some development of a new primary. But to test, even in principle, an existing secondary with a new primary for a new nuclear weapon Russia would have to conduct further tests well above any practical evasion threshold. Even then, however, since its performance would be similar to weapons already available, and Russia already has plenty of heavy-lift capability, the direct impact on U.S. security interests would be minimal.

If Russia does not have them available already, it could fully develop (if evasion were successful) light tactical weapons of yield of 1 to 2 kt or less. At the lower end of the very-low-yield category, Russia could develop and test new very-low-yield tactical weapons in the range of 10 to 100 tons. With respect to seismic detection, the 10-ton weapon could confidently be adequately tested under decoupling conditions even at Novaya Zemlya, and might even be tested in a steel or composite containment so that it would give no ground-shock at all. Indeed, with its experience in testing and weapon design, Russia could develop a 10-ton nuclear weapon using only hydronuclear tests in the kilogram-yield range, and be reasonably confident of its performance. Russia might even aim for a 10-ton weapon as a modification of an existing weapon of higher yield, of which it has a surplus. The United States would not be affected by Russia's conversion of, say, a 300-ton weapon of a type that it has tested a number of times, to a 10-ton weapon.

In summary, Russia's nuclear threat to the United States would not be significantly changed by successful evasive testing, but widespread speculation that evasive testing is possible might have a marginal effect on the nuclear-weapon incentives for neighboring states.

---

[6] JASON Report JSR-95-320, *Nuclear Testing - Summary and Conclusions* (McLean, VA: Mitre Corporation, August 3, 1995).

China

Given the modest current size and capabilities of China's nuclear forces overall—including the small number of strategic nuclear delivery vehicles (reported to be about 20 ICBMs at fixed sites)—it is not difficult to imagine changes in the numbers or character of these forces that would arouse U.S. security concerns. Such changes might include transformation of the strategic force to one based on mobile ICBMs, which might or might not be MIRVed, and the deployment of additional nuclear weapons for nominally non-strategic roles—such as short-range ship-based ballistic missiles or cruise missiles—that would have significant strategic as well as regional potential.

The basis of China's strategic nuclear posture with respect to the United States appears to be to hold a small fraction of the U.S. population at risk of nuclear attack. Given this approach, China has little incentive under present circumstances to use its missile payload capacity for multiple re-entry vehicles or multiple independently targetable re-entry vehicles—an approach that would lower the total deliverable yield. A U.S. national missile defense (NMD), on the other hand, could provide an incentive for China to MIRV as one means of improving penetration or of overwhelming a small NMD. Under those circumstances, China would have an incentive to increase substantially the number of delivered warheads. In that case, China might want to reduce the amount of plutonium or U-235 needed in a typical weapon and thereby allow a larger number of nuclear weapons to be built from a given amount of material. If China is limited instead by missile capacity rather than by its stock of fissionable material, weapons of smaller size and weight might be the goal, even if each used more nuclear material. The extent to which such developments might depend on further nuclear testing by China is not entirely clear. For example, in view of China's signing of the CTBT in 1996, it is likely that it had by then tested a strategic nuclear warhead suitable for its mobile intermediate-range ballistic missile now in flight test, or for an ICBM variant of it.[7] Further testing might bring additional improvements in yield-to-weight ratio and/or efficiency of utilization of nuclear material that China would find useful in the context of an effort to modernize and/or greatly expand its nuclear forces, if a decision were taken to do that. The indirect impact of China's testing could be substantial, in view of its influence on nuclear developments in India, Pakistan, and perhaps Japan. But it is quite clear that China could also achieve substantial increases in the capabilities of its deployed nuclear forces, if it wants to do so, without developing nuclear-weapon types beyond those it has already tested.

- Without a CTBT

In the absence of a CTBT, China would be able to conduct nuclear tests in the yield range needed to develop a more nearly optimum (lighter weight and perhaps more efficient use of fissile material) warhead for its mobile ICBM. But reduced size and weight depend as much or more on advances in non-nuclear aspects of the warhead (such as power supplies, and arming, fuzing, and firing systems) as they do on the nuclear package, so some improvements in this direction could presumably be achieved by China even in the absence of further nuclear testing.

China has far too few strategic weapons to attack a significant fraction of U.S. ICBM silos, and the United States has other more survivable strategic systems. With no option of a counterforce strategy against the United States within reach, China has little incentive to undertake the difficult task of developing a nuclear warhead resembling the U.S. W88, the special fea-

---

[7] NIC, September 1999, *op. cit.*

ture of which is its adaptability to a slim re-entry vehicle that can be effective against hardened point targets. In any case, a small force of warheads of this type would pose no greater threat to U.S. security interests than would a similar number of warheads of types the Chinese have already tested.

- Chinese advances strictly respecting a CTBT

Within a CTBT, China might exploit the possibilities for further developing its design expertise without nuclear testing, with an eye to the possibility of eventual collapse of the CTBT regime. China can certainly be expected to continue its stockpile stewardship program—smaller than that of the United States—including hydrodynamic tests with flash radiography and sub-critical tests. These activities will not impair U.S. security interests except in the unlikely event that they excited such Chinese interest in modifying an existing nuclear-weapon design or developing a new one that China elected to break out of the CTBT.

- Evasive tests within the CTBT

Overall, the direct impact of clandestine Chinese nuclear testing on U.S. security interests would be minimal. With the yields of concealable tests limited to 1-2 kilotons, China could not develop a new warhead. (Even if a test in excess of 10 kt could be fully decoupled by a factor 70, the decoupled signal would far exceed the seismic detection threshold at Lop Nor, and indeed at most other sites in China. At the few locations where this might not be the case, China would still need to worry about venting of radioactivity—which is very likely with decoupled nuclear explosions in hard rock—and about detection by U.S. and Russian National Technical Means.) As noted earlier, testing by China at lower yields for purposes of stockpile stewardship, if China decided that this was necessary, would not directly impact U.S. security interests. Of course, attempted clandestine tests that led to detection would indirectly impact U.S. security interests through the threat that this posed to the CTBT and non-proliferation regime.

India and Pakistan—States with Very Limited Test Experience

Neither India nor Pakistan is currently considered a strategic adversary of the United States, but the addition to the region of the kinds of nuclear-weapon capabilities these two countries have already demonstrated affects U.S. security interests at least indirectly, and further nuclear-weapon developments and deployments by these two countries would likewise be of concern to the United States.

- Without a CTBT (or without joining one)

Without a CTBT, India and Pakistan might undertake to perfect and modify their fission bombs and to develop thermonuclear weapons. This could result in a great increase in the destructive power of each weapon and at the same time provide an opportunity for increasing substantially the number of weapons that could be produced from a given amount of nuclear material. Ultimately, these countries could achieve strategic nuclear weapons that could be carried on considerably smaller ICBMs than would be needed for first-generation fission weapons.

More specifically, with a resumption of testing, India could refine its plutonium fission weapon. It could also master a thermonuclear weapon design with a size and mass compatible with delivery by its missiles. Such tests would with high probability impel Pakistan to resume its nuclear tests and might well provoke a return to testing by China.

With resumption of testing, Pakistan could further refine its enriched uranium implosion weapon and could develop boosted fission weapons. With plutonium available in the future from its new 50-70 megawatt reactor (producing some 15 kg of plutonium per year), Pakistan could also explore the plutonium implosion route, almost certainly with boosting. Following India, it could also attempt to design and test thermonuclear weapons. In this activity, and in obtaining maximum information from a test program, Pakistan would have much to gain from technology transfer from a Nuclear-Weapon State, if that were forthcoming.

- Achievements strictly respecting a CTBT

Without further testing, India could reproduce and stockpile the fission weapon it has already tested. It could also pursue a program of hydrodynamic and other subcritical testing to improve its understanding of plutonium properties and other aspects of weapon physics that are accessible short of criticality. Substantial doubt has been expressed about the validity of Indian claims to having tested a true thermonuclear weapon, and essentially no progress could be made toward stockpiling such weapons without violating a CTBT.

Pakistan similarly could manufacture and stockpile its enriched uranium fission weapons without further testing, and it could make progress toward a plutonium implosion weapon (perhaps even producing and stockpiling one of simple—and inefficient—design, in which it could have some confidence). Also like India, Pakistan could conduct hydrodynamic and other subcritical tests to improve its nuclear-weapon-related knowledge base.

- Evasive tests within the CTBT

As a large country with varied terrain, India might be able to avoid detection of a decoupled explosion up to perhaps 1 kiloton yield. It would have to guard against leakage of radioactive gas or particulates, however. The size of the country adds to the difficulty of detecting leaked radioactivity because of the delay before the released material is blown across a national boundary, but still India could not be certain of escaping discovery. As the table presented earlier indicates, clandestine testing in this size range would permit India to conduct one-point safety tests and, with difficulty, proof tests of weapons with yields up to 1-2 kilotons. But it would not suffice to develop boosted fission weapons or thermonuclear weapons. The conclusions about clandestine testing are the same for Pakistan as for India, but with less chance of successful decoupling even at yields below 1 kiloton, because the territory of Pakistan is much smaller. (Nuclear tests would perforce be closer to borders than need be the case for India, with greater likelihood of cross-border detection of seismic signals or test debris.) Testing by India or Pakistan could make a much greater relative improvement in its nuclear weapons than would a similar number of additional tests by China or Russia.

States of Concern Without Nuclear Test Experience

As noted earlier, with no nuclear testing at all it would be possible for would-be proliferant states to develop U-235 gun-type and simple plutonium or U-235 implosion weapons in which they could have reasonable confidence. But without nuclear testing, such states would not be able to improve upon the low material-utilization efficiency and poor yield-to-weight ratios of these first-generation weapons. Technology transfer from Nuclear-Weapon States could ease the transition to lighter and more efficient weapons, but for the recipient to have confidence in the performance or even the workability of its weapons, tests would be required.

To get the large efficiency gains and weight reductions associated with boosting, we believe an inexperienced state would need to test repeatedly at yields well above a kiloton, which it would not be able to conceal reliably. Tests in the 1 to 2-kt range, on the other hand, which could conceivably be concealed, might enable development—with difficulty—of an unboosted fission weapon in the 10 to 20-kt range that would be somewhat lighter than a first-generation "solid pack" implosion weapon. (This would not be needed if clandestine delivery by ship, transport aircraft, or truck were contemplated, but it would expand options for delivery by missile or military aircraft. The resulting improvement in short-range delivery capability would have some direct impact on U.S. security interests, for example by complicating the planning of U.S. military operations in conflicts similar to the Gulf War.) Tests in this 1 to 2-kt range could also validate the performance of the tested item as a weapon designed for this low yield, but designing such weapons would be difficult for an inexperienced state.

Repeated tests even at the 1 to 2-kt level would carry considerable risk of detection—tantamount to certainty for certain states. If an inexperienced state wanted to reduce this risk to a significantly lower level, it would probably try to limit test yields to 100 tons or less.

North Korea

- Without a CTBT

As an NPT signatory, North Korea is prohibited from nuclear testing even in the absence of a CTBT. In addition, under the Agreed Framework of 1994, North Korea is to be supplied two large nuclear reactors for the generation of electrical energy, and a number of other inducements, in return for its giving up the production of plutonium for nuclear weapons. These benefits would be withdrawn if North Korea tested nuclear weapons.

If North Korea were willing to abrogate the NPT and to give up the benefits it is enjoying under the 1994 Agreed Framework, and if it could produce or otherwise acquire plutonium or U-235 in sufficient quantity, it could develop and test a first-generation implosion weapon of perhaps 20-30 kt. With substantially more plutonium than it is suspected to have diverted in the past from its reactors, North Korea could have a test program leading eventually to a thermonuclear weapon. The resulting reduction in plutonium or U-235 needed for a given yield would increase the number of weapons that could be made from a given stock of fissionable material, and the improved yield-to-weight ratio of thermonuclear weapons would allow a given amount of damage to be done by fewer or lighter long-range missiles. This would represent a significant direct threat to U.S. security interests, and the indirect effects through encouragement of acquisition of nuclear weapons by Japan and South Korea would also be large.

- Strictly respecting a CTBT

North Korea might carry on nuclear-weapon-related activities permitted by a CTBT, even if such did not respect its obligations as a Non-Nuclear-Weapon State under the NPT. Such activities might include sub-critical tests with plutonium and hydrodynamic tests with surrogate materials.

- Evasive tests within a CTBT

Since North Korea has no suitable salt deposits, it could seek to test evasively in an underground cavity in hard rock. Such a test is very likely to leak radioactive particulates and gases, posing the risk of detection by the IMS or by U.S. NTM. Even in the absence of such leakage, a fully decoupled underground test of a 1-kt weapon would still provide an approximately 15-ton-equivalent seismic signal, which could readily be detected and located in North Korea by seismic means by a reasonably augmented IMS.

At the considerably lower yield that would stand a reasonable chance of evading detection, North Korea might test an unboosted implosion weapon leading toward a design that would give a yield of a few kilotons. But at lower yields more tests are required in order to provide equivalent confidence in the results.

Iraq and Iran

The acquisition of stockpiles of nuclear weapons by either of these states would have major political impact in the Middle East. Israel would believe its security and indeed its survival threatened. The uneasy relationship between Iran and Iraq would be destabilized. Such an event would likely lead to the acquisition of nuclear weapons by the other state, with possible assistance from outside powers. It could also lead to preemptive moves by Israel.

- Without a CTBT

Prior to the 1991 Gulf War, Iraq (despite being a party to the NPT) had mounted a large and varied clandestine program to acquire highly enriched uranium and possessed crude designs for nuclear weapons. Little remains of that program; it was dismantled under UNSCOM and according to the International Atomic Energy Agency apparently has not been reconstituted. Although Iran is also a member of the NPT, the U.S. government has stated that Iran is pursuing a nuclear weapons program.

Iran and Iraq were engaged in open warfare just over a decade ago, and the existing short-range missiles that both possess would be far more threatening to each other and to nearby states if these missiles carried nuclear weapons. Without a CTBT, either state might make the strategic calculation that its interests would be served by the acquisition of nuclear weapons and the demonstration of a nuclear capability. An underground nuclear test program might result (or even tests within the atmosphere), with eventual progression to boosted fission weapons and thermonuclear designs—providing a potential nuclear threat to U.S. cities, whether by ICBM, or by missiles of shorter range launched from ships, or by aircraft, or by detonation in a U.S. harbor.

- Strict Observance of a CTBT

If they could acquire the necessary nuclear material, Iran and Iraq could develop and produce—without nuclear testing—heavy and inefficient first-generation fission weapons. (This of course would violate their obligations as parties to the NPT.) But they could not improve on the material-utilization efficiency or the weight of these weapons without testing.

- Evasive testing under a CTBT

The yield of coupled underground tests in Iran or Iraq that might evade detection by the IMS or by United States Atomic Energy Detection System is somewhat higher than in North Korea. This arises because Iran and even Iraq are larger than North Korea, and seismic detectors are accordingly farther from potential test sites; seismic waves suffer more attenuation in this area because of the nature of the geology; and there are frequent earthquakes that raise the noise level and can be confused with underground explosions. For evasive testing, Iran and Iraq have salt deposits—unlike North Korea. As indicated in Chapter 2, the IMS primary network provides a detection capability at magnitude 3.25 in this part of the world, which would drop to 2.75 with the inclusion of the IMS secondary stations. The corresponding yields are some 0.060 and 0.020 kt, respectively, for a tamped explosion. As regards seismic detection, if Iran or Iraq could manage to construct a decoupling cavity in salt, it might attempt to test evasively at a yield up to 1 to 2 kt. The implications of this for nuclear-weapon development would be similar to the case of North Korea—namely, the possibility of progress toward moderate improvements in weight and materials-utilization efficiency compared to first-generation fission weapons, and the possibility of proof testing a low-yield fission weapon, but no possibility of achieving the much larger efficiency gains and weight reductions associated with boosting and thermonuclear weapons.

Summary of Potential Effects of Clandestine Foreign Testing

States with extensive prior test experience are the ones most likely to be able to get away with any substantial degree of clandestine testing, and they are also the ones most able to benefit technically from clandestine testing under the severe constraints that the monitoring system will impose. But the only states in this category that are of possible security concern to the United States are Russia and China. As already noted, the threats these countries can pose to U.S. interests with the types of nuclear weapons they already have tested are large. What they could achieve with the very limited nuclear testing they could plausibly conceal would not add significantly to this.

If Russia or China were to test clandestinely, within the limits imposed by the monitoring system, because they thought they needed to do so to maintain the safety or reliability of their enduring stockpiles, this would not add to the threat they would have posed to the United States in the circumstance that they were able to maintain the safety and reliability of their stockpiles without testing. Clandestine testing by Russia or China to maintain their confidence in their stockpile—although in violation of the CTBT, threatening to the non-proliferation regime, and not to be condoned—might actually be less threatening to the United States than either their losing confidence in the reliability of their weapons and building up the size of their arsenal to compensate, or their openly abrogating a CTBT in order to conduct the testing they thought necessary to maintain or modernize their stockpiles.

U.S. security could reasonably be judged to be directly threatened by clandestine Russian and Chinese testing for stockpile reliability only if the Russians and Chinese were able to maintain the reliability of their stockpiles by means of this cheating while the United States, scrupulously adhering to the CTBT, was unable to maintain the reliability of its own stockpile. This is precisely what has been hypothesized by some critics of the CTBT, but (as explained in Chapter 1) we judge that the United States has the technical capabilities to maintain the reliability of its existing stockpile without testing. If really serious reliability problems that only could be resolved through testing did materialize in the Russian or Chinese arsenal, moreover, it is unlikely that the degree of testing needed to resolve them could be successfully concealed.

In contrast to the cases of Russia or China, where their substantial prior experience with testing makes it at least plausible that they might be able to conceal some substantial degree of testing at yields below the threshold of detection, states with lesser prior test experience and/or design sophistication are much less likely to succeed in concealing significant tests. This is in part because of the importance of test experience in constructing cavities that can achieve seismic decoupling without leaking radioactivity, and in part because considerable weapon-design experience is required to achieve low yields. Countries with lesser prior test experience and/or design sophistication would also lack the sophisticated test-related expertise to extract much value from such very-low-yield tests as they might be able to conceal. They could lay some useful groundwork for a subsequent open test program in the event that they left the CTBT regime or it collapsed, but they would not be able to cross any of the thresholds in nuclear-weapons development that would matter in terms of the threat they could pose to the United States.

Undetected evasive testing under a CTBT would be limited to a level of about 1 to 2 kilotons and probably would be much less due to difficulties involved in evasive testing, particularly for states without extensive nuclear testing experience and availability of the required geological formations. They should properly be concerned that the yield of the test device might exceed that planned. To avoid this, an evader might conduct a series of "creep up" tests—which would increase the probability of detection and would be costly in terms of nuclear materials. The inability to test at yields above 1 to 2 kt would preclude the demonstration of boosted fission weapons, of primary nuclear explosives for driving the thermonuclear secondaries of strategic weapons, and the demonstration of thermonuclear weaponry. Possible evasive hydronuclear tests, which might escape detection by seismic means, would serve primarily to determine whether nuclear weapons are safe against accidental detonation at a single point; such tests in violation of the CTBT would not impair U.S. security interests and they would be costly in terms of the expenditure of plutonium.

In relation to two of the key "comparison" questions posed at the beginning of this chapter about the implications of potential clandestine testing, then, we conclude as follows:

- Very little of the benefit of a scrupulously observed CTBT regime would be lost in the case of clandestine testing within the considerable constraints imposed by the available monitoring capabilities. Those countries that are best able to successfully conduct such clandestine testing already possess advanced nuclear weapons of a number of types and could add little, with additional testing, to the threats they already pose or can pose to the United States. Countries of lesser nuclear test experience and/or design sophistication would be unable to conceal tests in the numbers and yields required to master nuclear weapons more advanced than the ones they could develop and deploy without any testing at all.
- The worst-case scenario under a no-CTBT regime poses far bigger threats to U.S. security interests—sophisticated nuclear weapons in the hands of many more adversaries—than the worst-case scenario of clandestine testing in a CTBT regime, within the constraints posed by the monitoring system.

Appendix A

Biographical Sketches of Committee Members

JOHN P. HOLDREN (NAS, NAE), Chair, is the Teresa and John Heinz Professor of Environmental Policy and Director of the Program on Science, Technology, and Public Policy in the John F. Kennedy School of Government, as well as Professor of Environmental Science and Public Policy in the Department of Earth and Planetary Sciences, at Harvard University. Trained in aeronautics/astronautics and plasma physics at MIT and Stanford, he previously co-founded and co-led for 23 years the campus-wide interdisciplinary graduate degree program in energy and resources at the University of California, Berkeley. He was employed as a plasma physicist at the Lawrence Livermore National Laboratory from 1970 until 1972, and has been a consultant to that laboratory continuously since that time. He is a member of the National Academy of Sciences (NAS) and the National Academy of Engineering (NAE), and he chairs the NAS Committee on International Security and Arms Control (which has conducted recent major studies of the future of U.S. nuclear weapons policy and of the management of surplus nuclear-weapon materials) and the NAS/NAE Committee on US/India Cooperation on Energy. He was a member of President Clinton's Committee of Advisors on Science and Technology (PCAST) from 1994 to 2001 and chaired PCAST reports on protection of nuclear materials in Russia, the U.S. fusion-energy R&D program, U.S. energy R&D strategy, and international cooperation on energy.

HAROLD M. AGNEW (NAS, NAE) is former President of General Atomics, and a former Director of Los Alamos National Laboratory. He is well-known for his pioneering contributions to weapons engineering. He has participated in numerous advisory capacities to such groups as the U.S. Arms Control and Disarmament Agency, the Council on Foreign Relations, and the White House Science Council. He was Chairman of the General Advisory Committee to the Arms Control and Disarmament Agency during the Nixon administration and a member during the Carter administration. Dr. Agnew served two terms as a New Mexico State Senator. He is a fellow of the American Association for the Advancement of Science and the American Physical Society, and is the recipient of the Ernest Orlando Lawrence Award and the Enrico Fermi Award from the Department of Energy.

RICHARD L. GARWIN (NAS, NAE, IOM) is the Phillip D. Reed Senior Fellow for Science and Technology at the Council on Foreign Relations, New York, and an emeritus Fellow at the

T.J. Watson Research Center of IBM. His expertise in experimental and computation physics includes contributions to nuclear weapons design, instruments and electronics for nuclear and low-temperature physics, computer elements and systems, superconducting devices, communications systems, behavior of solid helium, and detection of gravitational radiation. He was a member of the President's Science Advisory Committee from 1962-65 and 1969-72, and of the Defense Science Board from 1966-69. He currently consults for the Los Alamos National Laboratory, Sandia National Laboratories, and the Council on Foreign Relations, and is an active member of the JASONs. In 1998, he was a member of the 9-person Rumsfeld Commission to Assess the Ballistic Missile Threat to the United States. He has written extensively on nuclear-weapons-related issues over the course of several decades, particularly on the question of maintaining the nuclear stockpile under a comprehensive test ban regime. He chaired the State Department's Arms Control and Nonproliferation Advisory Board from 1993 to August, 2001. He is a fellow of the American Physical Society and the American Academy of Arts and Sciences, and a member of the American Philosophical Society.

RAYMOND JEANLOZ is a Professor in Earth and Planetary Science and in Astronomy, and is Executive Director of the Miller Institute for Basic Research in Science at the University of California at Berkeley. His expertise is in the properties of materials at high pressures and temperatures and in the nature of planetary interiors, for which he received a MacArthur Award. He serves on the University of California's President's Council and National Security Panel, is a member of the National Nuclear Security Administration Advisory Committee and of JASON, and chairs the NRC Board on Earth Sciences and Resources. He is a fellow of the American Academy of Arts and Sciences, the American Association for the Advancement of Science, and the American Geophysical Union.

SPURGEON M. KEENY, JR. is Senior Fellow, National Academy of Sciences. He served as an Air Force officer and civilian in the Directorate of Intelligence, HQ, USAF (1948-1954) in charge of the Special Weapons Section, responsible for intelligence on the Soviet nuclear weapons program and represented the Air Force on the Joint Atomic Energy Intelligence Committee, which produced national estimates of Soviet capabilities. Starting in 1956, he was in charge of the Atomic Energy Division, Office of the Assistant Secretary for Research and Engineering, Department of Defense, and was a member of the Gaither Security Resources Panel. In 1958, he was a delegate to the Geneva Conference of Experts on Nuclear Test Detection and subsequently to the negotiations on the Discontinuance of Nuclear Weapon Tests (1958-1960). From 1958 to 1969 he served as the technical assistant to the President's Science Advisor and concurrently from 1963 to 1969 as a senior member of the National Security Council staff, responsible for arms control and nuclear programs and policy. In 1965, he was staff director of the (Gilpatric) Presidential Committee on Nuclear Proliferation. From 1969 to 1973, he was Assistant Director of the U.S. Arms Control and Disarmament Agency (ACDA), responsible for backstopping the SALT I negotiations, and from 1973 to 1977 he was Director, Policy and Program Development, MITRE Corporation and Chairman of the Nuclear Energy Policy Study Group which produced the report *Nuclear Power Issues and Choices*. From 1977 to 1981 he was Deputy Director of ACDA, responsible for backstopping the SALT II and Comprehensive Test Ban negotiations. From 1981 to 1985 he was the NAS scholar-in-residence. From 1985 to 2001 he was President and Executive Director of the Arms Control Association. He is a fellow of the American Academy of Arts and Sciences and the American Physical Society, and a member of the Council on Foreign Relations.

ADMIRAL CHARLES LARSON (USN, ret.) is a graduate of the U.S. Naval Academy. During his naval career he was a nuclear submarine commander and commander of submarine forces, served two tours as Superintendent of the U.S. Naval Academy (1983-86; 1994-98), was commander of the U.S. Pacific Fleet (190-91), and Commander in Chief of the unified U.S. Pacific Command (1991-94). He has been involved in arms control and nuclear weapons policy issues as a Flag Officer (Admiral) from 1979 to the present time (SALT to START lll). He is currently a senior fellow at the Center for Naval Analyses.

ALBERT NARATH (NAE) retired in 1998 from his position as President and Chief Operating Officer, Energy and Environment Sector, Lockheed Martin Corporation. Prior to assuming that position, he was President of the Sandia Corporation and Director of the Sandia National Laboratories. Narath's active interests include federal science and technology policy, nuclear fuel cycle safety and environmental issues, and national security and arms control. He is a member of the National Academy of Engineering and has served on numerous NRC committees, including the Materials Science and Engineering Steering Committee and the Solid State Sciences Committee. He was appointed as a member of CISAC in 1998. Narath chaired, among other studies and committees, DOE's Fundamental Classification Review Group, the Oak Ridge National Laboratory's Advisory Board, and Lawrence Berkeley Laboratory's Advanced Light Source Program Policy Committee. He is a Fellow of the American Physical Society and the American Association for the Advancement of Science.

WOLFGANG K.H. (PIEF) PANOFSKY (NAS) is Professor and Director Emeritus at the Stanford Linear Accelerator Center (SLAC) of Stanford University. He served as Director of SLAC during the period 1961-1984. His field of expertise is experimental high-energy physics. He was a member of the President's Science Advisory Committee under Presidents Eisenhower and Kennedy, and the General Advisory Committee on Arms Control to the President under President Carter. He has also served on the NRC Committee to Provide Interim Oversight of DOE Nuclear Weapons Complex, the DOE Panel on Nuclear Warhead Dismantlement and Special Nuclear Materials Controls, and the NRC Committee on Declassification of Information for the Department of Energy's Environmental Remediation and Related Programs. He has served on ad hoc committees reviewing the directors of the DOE weapons laboratories for the University of California. He consults for Lawrence Livermore Laboratory, and is a current member of the JASONs. He was the Chairman of the Comprehensive Review Committee of the R&D Program of NN of the DOE. He is a current member and former Chairman of the NAS's Committee for International Security and Arms Control (CISAC) and chaired its study on management and disposition of excess plutonium.

PAUL G. RICHARDS is Mellon Professor of the Natural Sciences, Lamont-Doherty Earth Observatory of Columbia University, and former chairman of Geological Sciences at the university. He served on the NRC Panels on Seismological Data and Research Requirements for a Comprehensive Test Ban Treaty (1994-1995; 1995-1997) and was a member of the NRC Committee on Seismology. He was a Foster Fellow at the U.S. Arms Control and Disarmament Agency (1984-85 and 1993-94) and served as a member of the Red Team advising the Agency on the capability of the U.S. to verify compliance with the Comprehensive Test Ban Treaty (1997-1998). He spent a sabbatical at Lawrence Livermore National Laboratory (1989-90). During the CTBT negotiations in Geneva he presented an experts paper (1994) for the U.S. on the problems posed by chemical explosions. He is a fellow of the American Geophysical Union and former president of its seismology section. Dr. Richards is currently a member of the Board of the Seismological Society of America.

SEYMOUR SACK retired in 1990 after a career at the Lawrence Livermore National Laboratory, where he still consults. In his career he specialized in nuclear weapons design. He played a major role in the design of the "physics package" of many of the nuclear weapons in the current U.S. arsenal, and managed the development of two weapons systems still in the current stockpile. He participated in the JASON 1995 study on maintaining the nuclear stockpile under a comprehensive test ban. He is a recipient of the E. O. Lawrence Award (1973) from DOE and the Fleet Ballistic Missile Achievement Award from the U.S. Navy's Strategic Systems Program (1997).

ALVIN W. TRIVELPIECE (NAE) is a consultant at Sandia National Laboratories. Previously he was President, Lockheed Martin Energy Research Corporation and Director, Oak Ridge National Laboratory (1989 - 2000). He was Executive Director of the AAAS (1987 -1988), and the Director (1981 - 1987) of the U.S. Department of Energy Office of Energy Research (now Science). He was the Head-of-Delegation to the Peaceful Uses of Atomic Energy meeting between the United States and the Soviet Union (1986). He served as Chairman of the NAS Mathematical Science Education Board (1990 - 1995) and has served on numerous NRC committees. His formal training is in electrical engineering and physics. His research specialties include plasma physics, microwave devices, and particle accelerators.

Appendix B

List of Committee Meetings and Briefings

July 7, 2000 Washington, DC
Organization of the committee; no briefings

July 26-27, 2000 Washington, DC
Ambassador James Goodby, Department of State
John Parmentola, Defense Threat Reduction Agency

August 10-11, 2000 Washington, DC (Verification Subgroup)
Dr. William Leith, U.S. Geological Survey
Dr. Lynn Sykes, Columbia University
Dr. Larry S. Turnbull, Jr., Arms Control Intelligence Staff
Dr. Frederick Schult, AFTAC
Dr. Don Linger, Defense Threat Reduction Agency
Dr. Ralph Alewine III, Department of Defense

August 22-23, 2000 Lawrence Livermore National Laboratory
John Browne, Los Alamos National Laboratory
C. Paul Robinson, Sandia National Laboratories
Bruce Tarter, Lawrence Livermore National Laboratory
Roy Schwitters, University of Texas
John Parmentola, Defense Threat Reduction Agency
Victor Utgoff, Institute for Defense Analyses
James Silk, Institute for Defense Analyses
Brad Roberts, Institute for Defense Analyses
Bill Scott, SAIC
Bob North, SAIC
Charlie Meade, RAND

September 7-8, 2000 Washington, DC
David Crandall, Department of Energy
David Beck, Department of Energy

General John Shalikashvili and Ambassador James Goodby, Department of State
Dorothy Donnelly, Department of Energy
David Watkins, Los Alamos National Laboratory
Paul Brown, Lawrence Livermore National Laboratory
Gerald Kiernan, Department of Energy

September 21-22, 2000   Lawrence Livermore National Laboratory
Discussion of draft report; no briefings

October 6-7, 2000   Washington, DC
Dr. Ted Hardebeck, STRATCOM

November 6-7, 2000   Washington, DC
Discussion of draft report; no briefings

December 11-12, 2000   Lawrence Livermore National Laboratory
Discussion of draft report; no briefings

September 21-22, 2001   Lawrence Livermore National Laboratory
Discussion of response to Academy review; no briefings